T0332892

Medical Ethics for Physicians-in-Training

Medical Ethics for Physicians-in-Training

Jay E. Kantor, Ph.D.

New York University
School of Medicine
and College of Arts and Sciences
New York, New York

Plenum Medical Book Company • New York and London

Library of Congress Cataloging in Publication Data

Kantor, Jay E.
Medical ethics for physicians-in-training / Jay E. Kantor.
p. cm.
Bibliography: p.
Includes index.
ISBN 0-306-43194-7
1. Medical ethics. I. Title.
[DNLM: 1. Confidentiality. 2. Ethics, Medical. 3. Euthanasia. 4. Informed Consent. W 50 K165m]
R724.K3 1989
174′.2–dc19
DNLM/DLC 88-16272
for Library of Congress CIP

233 Spring Street, New York, N.Y. 10013

Plenum Medical Book Company is an imprint of Plenum Publishing Corporation

Printed in the United States of America

To VERONICA

Compañera de mi alma

Preface

The recent interest in biomedical ethics has resulted in the publication of a great many textbooks in the field. As good as many of these texts are, their attempts to encompass the ethical issues in all areas of health care have left them wanting in comprehensive treatments of specific areas that are of immediate concern to clinicians, and over-comprehensive in areas that are peripheral.

While the numerous anthologies of articles have the merit of not presenting students with a single biased approach, they usually have the disadvantage of presenting articles that are narrowly focused criticisms of other narrowly focused articles.

On the other hand, texts by single authors tend to be overly theoretical in their approach. The philosopher teaching ethics in a medical school or in a hospital setting must tread a difficult intellectual path. There are no "desert island" issues in clinical ethics, and few of the actual cases can be simply stripped down to clear conflicts between two philosophical theories. The horns of the dilemmas that he encounters are more likely to resemble a stag's horns than a bull's. A philosopher working in these settings must quickly change his accustomed approach to philosophical issues if he is to be effective. Very often he will be presented with an issue that he would prefer to mull over for a year or two, but which will require some sort of immediate direction or resolution because action must be taken.

Having worked in clinical settings for a number of years, I believe I have an idea of the issues that are most common and of most immediate concern to clinicians, and which will be of immediate concern to students when they begin their clinical work. Given that the amount of time allocated to training in medical ethics is usually (from the philosopher's point of view) inadequate, I have written a book that speaks only to those common and recurrent issues. Thus, for example, the reader will find little or nothing in this book that speaks to the large social issues about health care delivery, or to more specific issues about research and experimentation. Those absences are not meant to denigrate the importance of those and other issues.

Ethical analysis has come to have an enormous direct effect on medical policy and on laws relating to medical policy. Because of this, it is hard to write a book or article on medical ethical issues that will not be at least somewhat outdated by the time it is published. I have tried to avoid that problem by attempting to project a little into the future.

It is difficult to write a text such as this that is totally unbiased in its approach to the issues. I have tried to be objective, but readers are warned that my bias in favor of autonomy approaches to resolutions will probably surface frequently.

I would like to thank my students at the New York

University School of Medicine, my students at Methodist Hospital in Brooklyn, New York, and my colleagues and students at the Forensic Psychiatry Unit at Bellevue Hospital for their patience with a stranger in their midst. I would also like to thank my colleagues on the Ethics Committee at the Veterans Administration Hospital at Montrose, New York, for what they have taught me. I am grateful to Professor William Ruddick of the Philosophy Department of New York University for his philosophical input, and to the Humanities Council at New York University for its support of my teaching.

Finally, and most of all, I am thankful to Veronica Hinton for her patience and for her intellectual, editorial, and emotional support.

Jay E. Kantor

Contents

Medical Ethics for Physicians-in-Training

Chapter 1

Introduction and Philosophical Theories

THE NEED FOR TRAINING IN MEDICAL ETHICS

Whenever there is the possibility of interaction among persons, there are questions about how those persons *ought* to act.[1] The study of ethics may be described as the study of how persons ought to act. Since health care consists of interactions among persons, the practice of health care is inextricable from ethical issues.

LAW AND ETHICS

One common initial reaction to the perception of an ethical issue is to try to turn to the law for a solution. After all, the law always appears firm and unambiguous in its answers. And legal liability for malpractice seems to wait around every corner. Things are not that simple. First, laws regarding health care may and do change. We have seen important changes happen a number of times since the late 1970s. For example, the laws have changed and are changing in their criteria for the determination of death.

Second, the law may vary from jurisdiction to jurisdiction. Thus, at the time that this is being written, different states use different criteria to determine death. That means that a person who is legally dead in one state may be legally alive in another state.

Third, the law may be silent or fuzzy about issues. No law specifies which putative psychological therapies count as therapies and which count as quackeries. No law specifies precisely what is meant by "child abuse." Is child abuse to mean only beatings? Should it include neglecting to give sufficient hugs or refusing to help with homework?

Fourth, the law may require that a health care provider do something that is morally wrong. One of the important reasons for the interest in biomedical ethics over the second half of the twentieth century was the disclosures at the trials of the Nazi war criminals. There it was revealed that the law asked medical personnel to do numbers of things which, to dreadfully understate the case, were immoral. These included experimentation on persons without consent, medical examination of persons to determine whether they were fit for slave labor use or whether they were to be killed, and research into economical ways of killing large numbers of innocent persons. Among the conclusions drawn by the courts at Nuremberg was that sometimes persons have an overriding moral obligation to disobey the law.[1]

CODES OF ETHICS: THEIR LIMITS

Still another attempted approach to solving ethical problems is to turn to professional codes of ethics. But codes of ethics present difficulties too. For example, a code of ethics can prescribe policies or actions that may appear to be ethical precepts but, in reality, may have little to do with ethics per se. Thus the Hippocratic Oath directs the physician not to "use the knife" and not to give "abortion-causing drugs." While at least one of those directives has come down to us as an ethical precept, in fact both were probably intended to make sure that the Hippocratic physician did not demean his status by doing surgeon's or midwife's work.

Moreover, codes of ethics are often ambiguous. The Hippocratic Oath directs the physician "not to spread abroad" what he hears or sees in the course of treatment. The directive seems absolute, yet some would say that sometimes public interest and safety demand that the physician ignore the directive to keep his patient's confidences. Similar problems exist if we try to turn to more recent codes of ethics. As we shall see, they are often overly general, ambiguous, and liable to change.

Finally, there is the attempt to turn to religion for answers to the dilemmas. There are problems here too. Religious precepts differ from sect to sect, and differ in ways that are probably irreconcilable. We see this when we examine various religious positions on the abortion issue. We also see it when we look at the biblical injunction "Thou shalt not kill." The commandment has been taken by different sects as having different implications for issues such as abortion, euthanasia, and suicide. Moreover, there are ethical problems relating to religion within the practice of medicine. Religious freedom is taken to be a fundamental right in western constitutions. While that freedom often presents no problems to the physician, it does present a problem when we deal with sects that have religious beliefs that oppose standard medical practice. The classic instance of this is the Jehovah's Witnesses' refusal to accept whole blood transfusions.

Our belief is that the most rational and fruitful approach to attempting to solve ethical dilemmas is through an application of ethical analysis. We are of the belief that most ethical dilemmas arise because we are all pulled by a very few ethical theories which sometimes have, or seem to have, inconsistent implications about proper decisions. More often than not persons are not totally aware of the principles that are the source of their moral beliefs. At the very least, a knowledge of these principles will help to clarify thinking about issues. In turn, a clear view will sometimes help to dispel problems. What may appear to be an insoluble dilemma may in fact be soluble if thought through clearly.

Moreover, recent ethical analysis has begun to have a great effect on medical codes of ethics. In turn, codes of ethics help to set the "standard of practice" by which physicians are judged by their peers and by the courts. Medical ethics is now at the leading edge of the law and a study of medical ethics will help the medical student conceptualize what the law will be by the time he becomes a full-fledged physician.

It is well known that philosophy leaves many persons unsatisfied and with the feeling that they have left with more questions than they came in with. On the side of optimism, we will insist that many of the seeming insoluble dilemmas are, in fact, soluble. But the issues are difficult and difficult questions do not have quick and easy answers. Many of the solutions will be slow in coming and they will come over a tortuous and convoluted road. In some instances, the solving of dilemmas may take a time measured in decades rather than in hours. Issues usually interconnect in complex ways and, like discoveries in the sciences, the recognition of a hitherto unseen issue will usually open up a number of connected issues that require exploration. So, for example, the growing recognition of a patient's right to give or deny consent to treatment has opened up other issues that require analysis; what is to count as "informed" and what is to count as voluntary "consent"? It may be that some of the dilemmas have no solution; but that has never been proven and the issues are simply too important to give up the search.

We are also aware that each case that is encountered by the physician is both unique and complicated. No actual case is a "textbook case." But it would be a mistake to infer from the uniqueness of each case that general principles are inapplicable. Each patient is a unique medical case and presents unique problems in diagnosis, but that cannot make the physician throw up his hands in despair, throw away his text, and refuse to use general rules, principles, and criteria for diagnosis and treatment. Differences among cases are themselves explainable; they too can be subsumed under general explanatory principles.

The medical student who thinks that he can ignore training in ethics and simply deal with each case as it comes up is naive. To approach medicine that way will first leave him unaware of issues that may be present and, second, leave him picking answers out of a hat when he does spot problems. For the student who thinks that having good intentions and sensitivity is enough, we will issue a warning that possession of those qualities are a good first start towards being an "ethical" physician, but they are not sufficient. One of the reasons we have ethical issues to confront is that sensitive persons with good intentions disagree about just decisions.

Finally, no book or course in ethics can make persons act ethically. *Doing* the right thing is a matter of conscience and, sometimes, of courage. In many instances, the ways in which a medical student deals with patient care will be a function of the tone set by residents and attendings, and it would do them no harm to study this material. We can instill neither conscience nor courage here. Nor can we set the policy of a hospital or a service or unit of a hospital. What we can do is give the student some idea about where questions of obligation come up, and, we hope, some idea of what his obligations are. That is, we can help him *know* the right thing to do. The putting of theory into practice is up to him.

ETHICAL THEORIES

Those who teach ethics to clinicians sometimes observe an initial resistance from their students. Often, the source of the resistance is a belief that the study of ethics is futile because there are "no answers." When asked to give a coherent argument for that position, the cynics will usually come up with versions of one or the other of two[2] traditional meta-theories of ethics. These are "cultural ethical relativism" and "egoism." It is the author's belief that neither theory is ultimately supportable, and that neither one has much relevance to the actual trends in medical ethics. However, because belief in these theories is so widespread, something must be said about them. I shall say little here and thus, perhaps, be unfair to the more sophisticated versions of the theories. The reader can, if he wishes, look at the Bibliography for sources of more complete discussions.

Relativism

Some say that because moral values change from time to time and from culture to culture, it follows that there are no absolute "rights" or "wrongs." On closer inspection, this claim is usually seen to have two different components:

(1) The belief that moral values differ in different cultures. This belief is sometimes called "Descriptive Cultural Ethical Relativism." It is descriptive because it makes a claim about facts and does not prescribe what one ought to do.

(2) The belief that moral beliefs differ from culture to culture and that it follows from that fact that there are no absolute ethical truths. This belief is sometimes called "Prescriptive Cultural Ethical Relativism." It is prescriptive because it usually includes a claim that we ought to follow our culture's customs or beliefs when trying to decide moral issues.

Cultural Relativism

Cultural relativists make a *descriptive* claim. The claim is that anthropological evidence indicates that different cultures have different moral beliefs.[3]

However, one must be very careful about what is meant by saying that moral[4] beliefs vary from culture to culture. For example, it is known that Eskimos would take their old and put them away on ice floes to die. One might be tempted to say that this is a clear example of a moral policy that differs from present-day moral policies in developed nations. But suppose that the underlying moral reasoning of the Eskimos is the following: "The good of the tribe comes first. In the Arctic environment a whole tribe might die if it had to care for an old person who is unable to care for himself and unable to be of use to others." Under that plausible explanation, a seemingly strange custom becomes a not-so-strange decision derived from a belief in the moral principle of "the greatest good for the greatest number"[5] formulated under the particular circumstances that exist in the Arctic. Such a belief may, in fact, be completely analogous to the present-day policy in England that denies kidney dialysis to all those over 55 years old.

In fact, there do seem to be some moral beliefs that are universally held. For example, "'Unjustified killing of a person is wrong." Of course the criteria for what counts as "unjustified" and "person" may in fact vary from culture to culture.

In any event, those who believe that fundamental moral values differ from culture to culture must present more evidence than simple cursory descriptions of "strange" customs in various cultures.

Whether or not there are moral beliefs universal to all cultures, it is important to distinguish the *descriptive* version of Cultural Relativism from the *prescriptive* version, sometimes called Ethical Cultural Relativism.

Ethical Cultural Relativism

This theory holds that moral beliefs vary from culture to culture, and that it follows from that fact that there are no absolute moral truths.

There are major problems with this theory. First, the fact that different cultures have different moral beliefs does not logically entail that there are no moral absolutes. The cultures may each, severally, or all be mistaken about their beliefs. Certainly cultures can be mistaken about factual beliefs. For example, if a culture believes that malaria is directly caused by marsh gas, that culture is mistaken. Of course, some may say that there are absolute *scientific* truths, but that there are no absolute *moral* truths. However, that claim is unproven. Simply stating that there are no absolutes no more proves it than saying that there is no possible cure for the cold proves that there will be none.

Second, the difference between "scientific facts" and ethics may not be as great as it might seem. For example, sometimes a truth about medicine will or should lead to a change in moral beliefs. Thus, at one time it was believed that having epilepsy was a direct punishment for sin, and so the epileptic should be blamed for his "disease." The discovery of scientific facts about epilepsy did and should have changed moral beliefs about epileptics. The epileptic is no longer blamed for his disease.

Third, a consistent belief in cultural ethical relativism may have implications which many are not really willing to accept. For example, if each culture is its own measure of what is right and wrong, it may follow that there is no justification for criticizing *any* moral belief held by any culture. So, if a culture condones slavery, or demands punishment for having epilepsy, then that culture cannot be condemned. The physician confronting a member of a culture that believes prayer alone is the cure for a leukemia may have a hard time practicing medicine if he is both physician and cultural ethical relativist.

Moreover, cultural ethical relativism must be applicable to one's own society too. It is difficult to see how a belief in cultural ethical relativism would permit criticism or calls for change in one's own culture. In nineteenth-century America, for example, a cultural ethical relativist might have had to say that it was wrong to criticize the belief that epilepsy was due to sin.

Finally, the determination of a culture's ethics may be simple when the culture is isolated on an island. The task may not be simple in a complex culture. We are each members of many subcultures, some of which may have conflicting moral values. The "subculture" of a medical school may demand attendance at a class; the "subculture" of friends may demand attendance at a party at the same time. "Follow the dictates of your culture" offers no way to decide between these competing demands. The "subculture" of Jehovah's Witnesses may demand that they refuse whole blood transfusion for their children; the "subculture" of standard medical practice and law may demand that the children get blood.

While it may be possible to answer all these objections, the reader who finds ethical relativism convincing should be aware that they must be answered if he is to be consistent in his beliefs in cultural ethical relativism.

Self-Interest

There is another popular belief about the futility of attempting to do ethics. This is the claim that "People are only going to do what is best for themselves, so why bother to speak about 'ethics'?"

Here too the cynic is really making two mixed claims. First, there is the claim that persons cannot help but act for their own self-interest (selfishly). This is a belief about the nature of human psychology.[6] We will see later how Kant argues that the belief in inherent selfishness is untenable.

In a sense the belief in inherent selfishness is basic to both utilitarian theory and social contract theory. Yet both theories generate social ethics based on that belief.

In utilitarianism, we will see how individual interests are compounded to produce social policies and laws that take other persons' interests into account. We shall see how social contract theory derives a whole theory of rights and obligations from a belief in inherent selfishness.

Second, the belief in human selfishness might be taken to mean that people *can* (are able to) act for motives other than self-interest, but *won't* do so. That is, people are willfully selfish. The first claim is that people cannot help but act in their own self-interest; the second claim is that people can act otherwise but never do.

The belief that people *always* act for their self-interest seems empirically false. There do appear to be innumerable examples of persons who have acted against their self-interest for the benefit of others. As well, there seem to be many examples of persons who have acted because they believed the action was morally right, regardless of whether or not it was consistent with their self-interest. In the history of medicine, for example, there are researchers such as Reed who used themselves as research subjects in pursuing cures to very deadly diseases.

Those who are convinced that even seemingly altruistic actions are really selfish at base should carefully read the ethical theories that follow to see if Kant's objections to their beliefs can be refuted, or to see whether utilitarian and social contract theories change the implications of their beliefs.

> A 50-year-old man has been diagnosed with lung cancer. The cancer is incurable and advanced. His wife has been told the diagnosis and asks you not to tell him. You are examining the patient and he asks you, with apparent nervous bravado, "How am I doing?" Should you tell him the truth?

Mr. Jones, a 40-year-old unemployed electrician, needs a renal transplant. He has no children. He is unmarried. He has been told that a kidney has become available, and he is now in the process of checking into the hospital for the transplant surgery. While he is checking in, the transplant surgeon receives word that Dr. Smith is also in need of a kidney. Both Smith and Jones are suitable recipients for the kidney. Neither is suitable for dialysis. Dr. Smith is a well-known cardiac surgeon. He is in the midst of perfecting a technique for replacing very badly damaged hearts. So far, his technique has saved hundreds of lives. In addition, Dr. Smith is married and has three young children.

The transplant surgeon wonders whether he should tell Jones that a mistake was made and that the kidney is unsuitable, and then give the kidney to Smith.

These are not unusual examples of the types of ethical dilemmas that occur in medicine. Notice first that the dilemmas may *involve* questions about appropriate medical treatment, but have another dimension that has little to do directly with treatment. For example, the medical facts in the transplant example are quite clear. Both persons need the kidney. Both are medically suitable recipients for the available kidney. The problem here is ethical, and has to do with the just allocation of scarce medical resources.

The dilemma in the other case has a medical side; we wonder what the medical effect of telling the truth to the patient will be. However, the other side is ethical; we believe we have at least some moral obligation to tell the truth to the patient.

The ethical analysis in each case is complex, and actual cases can be even more complex. However, such an analysis is necessary if we want to try to approach a just decision. Many, if not all, of the cogent factors may be analyzed in terms of the theories that follow. If we look at the kidney allocation case, for example, we would have to consider at least the following factors:

1. Do we have a stronger obligation to give the kidney to the electrician because he is already under our care? (social contract theory)
2. Do we have a stronger obligation to give the kidney to the surgeon because he is of proven use to society? (utilitarian theory)
3. Do we have an equal obligation to both because they both are persons, regardless of their value to society? (Kantian/autonomy theory)

Our analysis may begin with these questions. We may find that although at first glance the theories seemed to entail different resolutions, a full analysis shows us that they all entail the same solution. On the other hand, it may very well be that a rigorous analysis still leaves us with a dilemma. In those instances, we will be forced to take a stand and choose our stance from whichever theory seems to have the best foundation. We will say a bit more about that at the end of this section.

Utilitarianism

"Good" = Situation with more existing happiness than any other possible situation would have had.

Right action = That action, policy, or law that produces greatest happiness (and/or least pain) for the greatest number of sentient beings affected by that action, policy, or law. Actions are not intrinsically right or wrong; they are judged according to their *consequences*.

Health = Continuing state of happiness.

The attempt to justify paternalistic treatment of competent persons is to be found in utilitarian theory. In addition, the theory gives a rationále for taking the public interest as primary and standing above individual interests.

Utilitarianism was developed in England and as a British theory it has certain points in common with Anglo-American approaches to the sciences. The Anglo-American approach to scientific explanation has tended towards empiricism. The major tenets of empiricism are: (1) a belief that knowledge is derived purely from experience and, connected to that, (2) a reluctance to accept the existence of entities that cannot be observed. This differs from the continental approach to scientific explanation—rationalism. Rationalists are more ready to accept both that there is knowledge that is not obtained from experience and that there may be entities that cannot be observed (e.g., "self," "will," and "ego"). The empiricist tradition has tended to accept the "tabula rasa" view of mind—that is, the claim that we are born with neither beliefs nor knowledge nor complex instincts, but instead that our beliefs are drawn from postnatal experience. We see Skinnerian theory as exemplar of the empiricist view and Freud and Lorenz as representative of rationalist views. Skinner attempts to explain actions in terms of observable and mostly learned behavior, Freud and Lorenz are more willing to explain behavior in terms of unobservable entities and, in part, inborn patterns.

In forming a theory of the good, the utilitarians turn

to an observation of human and animal behavior. From their observations they conclude that all motivated behavior has as its motivation a desire for pleasure or, given dismal alternatives, a desire to avoid pain. They further conclude that pleasure (and absence of pain) is the good. Therefore a *right* action, policy, or law is defined as an action, policy, or law that produces more pleasure (or less pain) than any alternative action, policy, or law would have produced. The moral goal is to maximize pleasure and, in the realm of social policy, just laws and policies are those that produce "the greatest happiness for the greatest number."

How does one judge which course of action will produce the greatest happiness? The utilitarians were attempting to create a science of ethics. Problems about the production of good are to be approached as if they were scientific-technological endeavors; the utilitarian gathers as much information as possible about the probable consequences of alternative choices, and then chooses the course of action that will most probably produce the greatest happiness. Application of such a theory is not simple. In terms of patient care, one has to try to predict not only the *immediate* effects of an action upon a patient and his relatives but also the effects of the putative action upon the staff and all other persons who will be affected. Moreover, since particular decisions sometimes have a way of turning into general policies, the utilitarian has to take into account the possibility that the decision will become a general policy, and try to determine the effects that policy would have upon the general happiness. Since any action has a ripple effect, such a determination is quite difficult and complicated.

Note that for the utilitarians there are no prima facie obligations to tell the truth to patients, to keep confidentiality, or to obtain an informed consent. Such obligations exist only if greater happiness will be produced by telling the truth, by keeping confidentiality, or by obtaining an informed consent. Correspondingly, there are no "absolute" rights to privacy or to autonomy. The *only* absolute duty there is to persons is to take their interests into account when making a decision. The only absolute right that persons have is to have their interests considered in the reckoning.

An example of a utilitarian reasoning is one justification sometimes given for the policy of confidentiality: "Persons have no inherent right to confidentiality. However, the general knowledge that confidences will be kept by mental health professionals will encourage people to seek treatment and further encourage them to be completely honest in treatment. That will promote the general welfare of society as well as make treatment more successful."

Paternalism

Utilitarian theory also provides the attempted justification for paternalism, which may be defined as a policy of doing something for someone's benefit without asking his permission, or contrary to his avowed wishes. The derivation of the word makes the analogy to parental treatment of children. Presumably, a parent may make decisions for his children without asking their permission. Paternalism takes two forms: *State paternalism,* where the government limits the liberty of individuals for their own sake or for the sake of the population at large. Example of state paternalism include: compulsory vaccination; the requirement for prescriptions for some drugs; F.D.A. drug-testing requirements;

requirements for the use of crash helmets; requirements of licensing and certification of physicians.

Individual paternalism is paternalistic treatment of individuals by other individuals. Examples include: withholding the truth from patients; lying to patients; the use of placebos in treatment; treatment of competent patients without their consent.

Kantian/Autonomy Theory

Good = Situation in which there is happiness without persons having been used to achieve that happiness.

Right action = An action that (1) could be generalized as a rule for all "humankind" to follow; and (2) does not use unconsenting persons as means to achieve some end. Actions are judged right or wrong according to *intent* rather than *consequences*.

Health = Condition of internal autonomy.

A twenty-five-year-old male enters a hospital as a service patient for incision and drainage of perianal abscesses.

The surgery is uneventful.

The next day the staff, consisting of residents and medical students, make rounds. They enter the patient's room. The patient is asleep. The resident wakes him by telling him to turn over. Without introductions, without asking how the patient is, they proceed to examine the anal region. The patient asks, "What is going on?" The surgeon, noticing an abscess missed in surgery ignores the patient and asks a medical student for a syringe and lidocaine.

> Without a word, the resident injects the patient in the area of the abscess. The patient squirms and shouts "Communicate with me doctor!" and "You are supposed to be a doctor, communicate with me!" The resident ignores him, gives him another injection of lidocaine, and drains and dresses the abscess. The staff leaves.

What is horrifying about this case? It is not a life-or-death situation. It is not a controversial no-code order, or a denial of treatment to a dying patient or to a newborn. The procedure was done, treatment was completed, and completed quickly. The medical students present learned something. The patient will be out of the hospital in a day or so.

Recent trends in dealing with ethical issues in medicine have stressed individual autonomy. Approaches to patient care have become quite Kantian.

Although Kant attempted to reconcile empiricist and rationalist approaches, his ethical theory falls more on the side of rationalism. He begins his most important ethical work, *The Foundations of the Metaphysics of Morals*,[7] with a critique of the pleasure/pain theories of the early utilitarians. He claims that while a pleasure/pain theory of motivation will work to adequately explain the behavior of nonhuman animals, it is not a satisfactory explanation of human behavior. Unlike animals, (which Kant believes are solely motivated by instinctual drives and the desire to seek pleasure and avoid pain), normal adult human beings have the ability to overcome instinctual drives. More precisely, we have the capability of acting or attempting to act for the reason that we believe that the action is the morally right thing to do. We can overcome instinct and desire because we each have a *rational will*[8] that enables us to act from motives of *obligation* or *duty*. While nonhuman animal behavior is bound by laws of nature (that is, their inherent biology), we have the ability to *legislate* our behavior—to govern ourselves. This *autonomy*, this ability to overcome desire and instinct and to be responsible for our actions, is what gives us inherent, intrinsic, and infinite worth. The ability to overcome natural inclinations, particularly self-interested ones, makes us worthy of respect and gives us dignity. In the Kantian view, the ability to legislate is what defines us as *persons* rather than as *things*. Things may be used as means or tools to achieve various goals. Persons may not be used solely as means. In Kant's words, "So act so as to treat humanity, whether in yourself or in others, always as an end within itself, never solely as a means."[9]

This communication between I.G. Farben Chemical Corporation and the Auschwitz concentration camp horrifyingly exemplifies the violation of autonomy theory:

> In contemplation of experiments with a new soporific drug, we would appreciate your procuring for us a number of women. . . . We received your answer but consider the price of 200 marks a woman excessive. . . . Received the order of 150 women. . . . The tests were made. All subjects died. We shall contact you shortly on the subject of a new load.[10]

Since we define our own nature, we cannot simply look to nature or science, as the utilitarians do, for a description of what are to count as right actions, policies, and laws. Instead, we must examine each possible action and see whether we could choose to generalize the action and set it into a law that (1) all persons *could* follow and that (2) we would *want* all persons to follow. Now, while many actions could be turned into universal laws that fit those criteria, the theory sets further strictures. The most important stricture for our purpose is the absolute prohibition against using persons solely as means to achieve some end. Such use is not permissible even if the intent is to benefit the person. In the latter case, we would be showing disrespect for the person's autonomy. For example, to lie to a competent patient "for his own benefit" is to show a disrespect for his autonomy. That would have the effect of infantilizing the person, saying by implication, "You are less than a responsible being and not capable of making decisions, and so we will make them for you."

Nor may we use persons as means to achieve happiness for others, or to achieve happiness for society at large. While the utilitarian may claim that it is a matter of empirical fact that using persons without their consent would never produce "the greatest happiness for the greatest number" in the long run, there is nothing *inherently* wrong with using persons in utilitarian theory. Kantian theory claims that we have strong duties to produce happiness for ourselves, for others, and for society in general, but happiness may never be achieved at the expense of violating a person's autonomy—his right to self-determination.

Moreover, the patient is to be treated (as is any person) with what Kant calls "practical love" rather than what he calls "pathological love." Practical love is a rational attitude towards persons that is exemplified in the injunction "Love thy neighbor as thyself." One should not let one's emotions overcome one's reason when treating patients. That is the case whether the physician personally finds the patient's personality despicable or whether he feels enormous empathy for the patient.

In one sense the very taking on of a person as "a patient" places him into a setting in which he is "used." That is, he is used in order to gain a livelihood, or to learn medicine. But the prohibition against using persons *solely* as a means permits that "use" with the provisos that: (1) The physician has the person's consent; and (2) his being a patient is a "'role," and that his primary being is as a person. That is, he is a person first and a patient only second. Thus, anything in the professional–patient relationship that would violate the primary person–person relationship is prohibited. For example, the fundamental right to self-determination cannot be bargained away in the physician–patient contract.

For another example, dealing with a patient as if he were merely a case number or an example of a disease would be wrong. Furthermore, it would be wrong even if the outcome of treatment were to be the same as it would have been if he had been treated as a person. Certainly, much of our horror about the handling of the case that opened this section can be explained as horror at a person being treated as if he were a thing or a disease.

Of course, each person doing "moral reckoning" is himself a person. Thus, I myself am a person like any other and I have a duty to treat myself as I would any other person. That implies that I have duties to myself. Not every ethical theory makes that claim. Some would claim, for example, that if my actions or behavior have no effects upon others, I am free to perform those actions or indulge in that behavior. The claim that *any* action or behavior that is agreed to by consenting adults is permissible is not one to which Kant would agree. Such behavior may demean the persons involved. Furthermore, since I am a person like any other, I ought not to allow myself to be used in ways that are demeaning. Thus, by implication, Kantian theory prohibits passivity. For Kant, the person who allows himself to be totally directed and controlled by others is behaving as less than a person, even if he is happier that way. We have duties to respect and make use of our own autonomy—to be responsible and "self-determining."

There are some less obvious but important implications of Kantian theory. First, the rights-status of those humans who are less than autonomous is in doubt. Kantian theory gives us little to go on in regard to the rights of infants, children, the retarded, and the mentally ill. Furthermore, Kantian theory gives us little to go on in attempting to decide who is less than autonomous. That is, he presents no real theory of mental illness or competence–neither diagnosis nor etiology. And those are crucial matters in issues such as involuntary commitment, the right to refuse treatment, and the insanity defense, for these very issues have to do with the patient's degree of internal autonomy. There are three problems here: (1) How do we distinguish these "patients" from normal autonomous patients? (2) Do these "patients" have rights at all? (3) If they have no rights, or fewer rights than full-fledged persons, what is the nature of our obligations towards them? The same problems exist for social contract theory.

Natural Law Theory

Good = individual and societal happiness achieved according to fulfillment of natural purpose (design, essence).

Right action = action in accord with essence, with intention to do what is right.

Health = condition of accord with one's essence.

Natural law theory is more than just an ethical theory; it is also a theory of science and medicine. The theory has had an enormous influence upon science, medicine, law, and religion. For example, clinical psychological theories were traditionally bound to natural law theory and are only now emerging from natural law presuppositions. Recent major shifts in psychiatric diagnosis reflect this change.

According to the theory, everything in the universe has an essence. "Essence" is interpreted as roughly equivalent to "design" or "purpose," or "ultimate nature." Each thing in the universe tends to move towards fulfilling its essence. Thus, for example, oak trees have an essence; the essence of an oak tree are those inherent characteristics that make it an oak tree and not an elm or a dog. Acorns have their essence and that is to grow into oak trees.

As usually interpreted by natural law theorists, humans have their essence too. We are the "rational animal." That means we have a physical being, a functioning body that must be taken care of. We also have our rational nature, and that includes our ability to control how we behave. That is, it includes our "free will." Our rational nature must also be kept functioning. Our possession of rationality means that (1) we can comprehend our purposes, and (2) we have the ability to choose or to refuse to go along with our purposes. That ability to choose can make us culpable for doing wrong. According to natural law, acting contrary to one's nature is wrong, irrational, and unhealthy. The three terms are almost synonomous for the theory. Moreover, we cannot achieve happiness unless we follow our nature.[11]

More specifically, natural law theorists have said such things as the following about human nature:

We have reproductive organs, the purpose of which

> is propagation of the species. Any other use would be a use contrary to our essence and hence unnatural. The pleasure attached to sex is designed to encourage the procreative act and not an end in itself. Unnatural uses of the reproductive organs would include recreational sex, masturbation, homosexuality, and any sexual act that doesn't terminate with ejaculation within the womb.
>
> Moreover, we are social creatures. That means, among other things, that we are to have families, but only within the context of heterosexual marriages.
>
> The natural end of the fertilized human egg is to develop into a person; to interfere with that process would be wrong.
>
> Women were designed to have children and tend the hearth. Any woman who shirks those natural functions is immoral and irrational.[12]
>
> We have an inborn desire to preserve our lives; therefore any desire to intentionally end one's life is irrational and wrong.

The past influence of natural law theory on conceptions of health should be apparent from some of the examples above. The remaining influences which may not be as apparent will be discussed later when we speak more directly about "health" and "illness," and when we speak about abortion.

Extraordinary Means

There are other parts of the theory that are extremely important for medicine:

1. It is part of the theory as usually interpreted that everything in the universe was put here for our use. That means that we can do whatever we please with the rest of nature (including animals). So, while we have a duty to make human fetuses into persons, we have no duty to turn acorns into oak trees. We can use acorns and oak trees as we please.
2. Extraordinary means to save or sustain life: Extraordinary means are, for natural law theory, means of treatment that are above and beyond the natural unfolding of events. They are means that would require an unusual effort or would be an unusual burden to attain or use. Thus, there is no absolute duty to use extraordinary means to keep the patient alive.[13] Neither is there an absolute duty of a patient to accept such means if they are offered.
3. Natural law theorists gave important analyses of the differences between "acting" (committing) and "refraining from acting" (omitting). One conclusion of the analyses is that, given a choice between *doing* wrong and *letting* a wrong occur, the lesser of the two evils is to let the wrong occur. That is, it is worse to *do* evil than it is to *let* an evil occur.
4. In line with (2) and (3), we have *perfect* duties and *imperfect* duties. Roughly, the distinction is that perfect duties are duties not to harm (duties of non-maleficence), while imperfect duties are duties to do good (duties of beneficence). Perfect duties have moral priority over imperfect duties. In terms of law, the theory would hold that a layman[14] can be penalized for harming other per-

sons, but cannot be punished for refraining to do good for other persons.

Social Contract Theory

Right action = Action permitted by the contract.

Good = Condition in which an individual's interests are satisfied.

Health = Condition in which one's powers are fully functioning and intact.

The last of the important major theories is social contract theory. Historically, different versions of social contract theory had a major influence upon law. Like natural law theory, from which it borrows certain concepts, it was used as a major theoretical foundation for the great eighteenth-century constitutions. For our purposes, it is most important for helping to understand and define the connection between rights and obligations. In particular, it has much to offer for untangling the issues of professional roles and professional obligations (e.g., physician–patient; physician–institution; physician–society; licensing, etc.).

The following is a synopsis of a version of social contract theory that was offered by Thomas Hobbes.[15]

Humans are born with a very few innate characteristics; we are self-interested (selfish), we are all approximately equal in our powers (no one is a superman), we have strong desires to preserve our lives and our powers.

Given those assumptions about human nature, Hobbes asks us to imagine an early presocietal existence. In such a "state of nature" there is "war of every man against every man."[16] Each person tries to get whatever he can for himself. He is free to kill, steal, and lie, and will do whatever he can in order to get what he wants.

But no one person is all-powerful. Each person is in constant danger of death or theft at the hands of others. Under such conditions no person can be secure in his possessions or his life. (In a state of nature "life is solitary, poor, nasty, brutish and short."[17]) One may gather goods to his cave. But one must eventually sleep, and while asleep, one can be murdered.

In order to gain security, persons begin making covenants (contracts) with their neighbors: "If you watch me and my cave while I sleep and promise not to kill or rob me, I will do the same for you."

Notice what happens as a result of this contract—I gain some *rights*—the right that you not kill me, the right that you not rob me, the right that you protect or warn me. In return, I incur certain *obligations*—the obligation not to kill you, to watch your cave, etc. Also my *liberty* is limited—I am no longer at liberty to try kill you, or to try to rob you. At the same time, you also gain those rights and incur those obligations. Since we can't trust each other to keep the contracts, we create a government that has the obligation of policing these contracts.[18] Since we are self-interested, we try to cede as little of our autonomy (incur as few obligations) as possible in entering into these contracts.

One of the ways that government enforces contracts is to place a high price upon breaking them. Hence social contract theory conceives of punishment as deterrence. Government deters certain acts by trying to make the probable losses for acting greater than the probable gains.

And so we have a society and so too the formation of rights. As well, we have the creation of obligations.

In the social contract view there are no rights without corresponding obligations. And, interestingly, we have an ethical system that does not depend on the existence of "good" people.[19] One may think of the definition of "person" given by Hobbes as very similar to that given by Kant.

While social contract theory has been used to justify minimalist government—the concept that the only function of government is to police and protect persons' freedom to *try* to get what they want—some have said that it could be used to justify a more active government. For example, if the population decides that their interests would be served by having a right to health care, then they may make that a part of the social contract.[21]

Rights

> We hold these truths to be self-evident: that all men are created equal and endowed by their Creator with certain unalienable rights. Among these are Life, Liberty, and the Pursuit of Happiness. . .

We probably seldom think much about those statements, but a cursory examination reveals them to be quite problematic:

1. In what sense is it "self-evident" that all men have rights?
2. What is meant by "men"? We know it is meant to include women, but does it also include fetuses? Convicted criminals? The comatose? The severely retarded? Persons in other countries?[22]
3. What is meant by "unalienable"? Does that mean these rights can *never* be taken away? Can they be voluntarily waived, or given up?

4. "Among these rights are . . ." What are the others?
5. Is the "right to life" simply a right not to be killed (a right to protection), or is it also a right to be kept alive if one is in danger of dying (a right to at least emergency health care)?

Substantial work has been done in the past 30 years in forming a coherent theory of rights. Theories about rights per se are all rather recent (of a few hundred years at most), and there is still a great deal of work to be done. One step that has been taken is towards a categorization of rights:

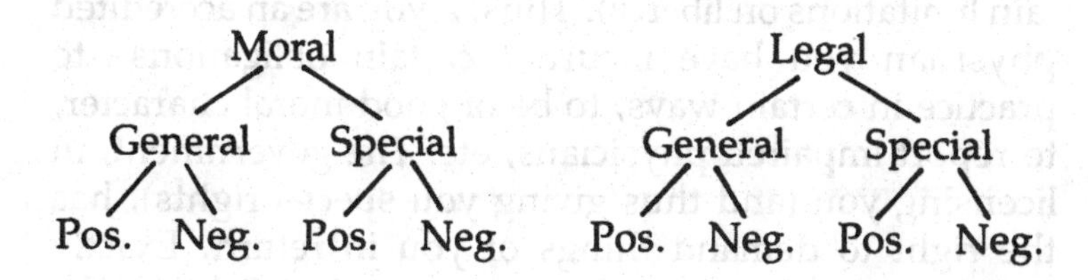

The broadest category difference is between Moral rights and Legal rights. A moral right is one that ought to be recognized by a legal system, but is not always so recognized. So, for example, blacks in the pre-Civil War South had a *moral* right to freedom even though they had no *legal* right.

A legal right is a right that is actually recognized in a legal system. There are legal rights that do not always correspond to moral rights. Thus, for example, whites had a legal right to own slaves in the pre-Civil War South, but not a moral right. A dictator or absolute monarch may have legal rights that excede his moral rights.

The next major category distinction is between *general* rights and *special* rights.

A general right is one held by all persons regardless

of circumstances. Thus, for example, all persons have rights to life, liberty, etc.

A special right is one that comes out of an agreement made between persons or groups of persons. For example, if you own land you may give me a (special) right to walk through your land. Or, if I am accepted as a member of a club, I may gain the right to use the club's facilities.

If I am a member of a health plan, or an HMO, I gain certain rights. If you are a licensed or certified physician you gain certain (special) rights, including, of course, the right to treat patients.

With special rights come obligations (which are certain limitations on liberty). Thus, if you are an accredited physician you have incurred certain obligations—to practice in certain ways, to be of good moral character, to report impaired physicians, etc. The government, in licensing you (and thus giving you special rights), has the right to demand things of you in return. Exactly which demands the government may justifiably make is problematic. For example, may it force you to cede the right to organize into unions? Demand that you help accident victims in the street? Demand you be of "good moral character" even if your immoral actions don't affect your skill or abilities as a physician?

These are problems in regard to the contract between physician and patient too. In the traditional view, the primary obligations incurred by the physician towards his patient were, first, to do no harm, and, second, to do good for his patient—to be *beneficent*. With the acceptance of autonomy theory approaches, the duty of beneficence is taking third place. The primary obligations are seen as first, to do no harm and, second, to recognize the patient's right to self-determination—even if the physician believes that the patient is making a choice that is contrary to the patient's own good.

Finally, there is an important distinction between *negative* rights (sometimes called "liberties" or "option rights") and *positive* rights (sometimes called "welfare rights" or [though this is weaker than a right] "entitlements").

A negative right is a right to try to get something, or to pursue something, or to hold on to something, without interference by others. Thus, for example, the right to life interpreted solely as a right not to be killed, or the right to try to find employment, or the right to free speech.

Historically, negative rights were the first to be recognized within legal systems. In conservative political theory,[23] they are the only rights that exist as general rights. From that conservative point of view, the only function of government is to protect persons' freedom to try to get what they can.

Positive rights are rights to be given something. So, a right to be given food if hungry, or a right to be kept alive if one is dying, or a right to health care, would all be positive rights.[24]

The claim that there are general positive rights is recent and controversial. Liberals will generally believe that there are positive rights; conservatives will not, though very few conservatives will believe there are *no* positive rights. (The very acceptance of a social "safety net" is a recognition of positive rights.) Note that, as of yet, the U.S. does not fully recognize that there is a general right to *any* health care under *any* (including life-death) circumstances.

Genesis of Rights

The following should give you an idea of the realation of rights to the various ethical theories.

Social Contract (Hobbesian). We enter into a community with other persons in order to protect ourselves and further our interests. In entering into the community we give up certain liberties (e.g., the "right" to kill) in order to gain security (e.g., to get a legal right not to be killed). We enter into "covenants" with others ("If you don't kill me, I won't kill you") and set up governments to enforce these contracts. Each agreement gives us a right and, at the same time, obligations. Breaking the covenant means forfeiting one's legal right—e.g., if you kill a person, you give up your legal right to life. In the social contract view of rights, all rights are really *special* rights. Moreover, only those capable of understanding these agreements have rights. That implies that fetuses, some of the psychotic, and some retarded do not have rights.

Presumably, a population can covenant for positive rights (e.g., public health care rights) as well as negative rights.[25]

Legal Positivism. This is the view that there are no rights or duties other than those recognized within a legal system. That is, there are only legal rights. After the Nuremburg trials of Nazi war criminals (where it was ruled that simply obeying laws is not enough but that there are higher moral laws), legal positivism is out of favor. Legal positivism ties in closely to moral relativism and is subject to the same criticisms.

Natural Law. Rights protect the achievement of humans' natural ends.

Utilitarianism. Not everyone believes that the concept of rights is compatible with utilitarian theory. If there are such rights it is because they are thought of as conducive to the production of the greatest happiness. Thus a utilitarian might believe that there is a right to life because people are happier knowing their lives are protected by such a right. However, utilitarian theory requires that if it can be shown that violation of a right will produce greatest happiness, then it is a duty to violate that right.

Kantian/Autonomy Theory. Rights are absolute protection against the use of "rational wills" as means (as things), rather than as persons.

APPLYING THEORY TO PRACTICE

We have said that few of the actual cases encountered in clinical work can be quickly, neatly, and simply analyzed as clear conflicts between the implications of the various ethical theories. Usually, it is the broader ethical issues that are more easily seen as conflicts between theoretical approaches. Thus, for example, the problems of informed consent may be easily discussed in terms of the conflict between autonomy and utilitarian approaches to the issue. On the other hand, a specific case in which informed consent is an issue will likely be complicated by the presence of other issues.

Individual cases often have many layers of conflict, if only because there are usually many persons involved in the case. The patient, his family, his friends, physicians, nurses, social workers, and hospital administrators may all be involved. They may have different perspectives, and there may be a number of types of obligations and rights involved in their interactions. In analyzing an actual case, it is crucial to get a full description of the relevant facts. That ordinarily will include a description of the different perspectives of the various persons involved. In the best of circumstances, conflict may simply be due to a misunderstanding about the facts.

Thus, for example, a nurse who has qualms about the validity of consent for a withdrawal of treatment order may simply not have been aware that the physician had fully discussed the order and its ramifications with the patient. Each may have had the same concern about the patient's right to self-determination; but the nurse may not have known that steps had been taken to ensure that the right was honored. Sometimes the resolution of a case may require that the physician or case presenter go back to get more information from the patient or from other members of the treatment team. The importance of good communication within the treatment team as well as between the treatment team and the patient is crucial, not only to good medical treatment as such, but also to ethical medical care.

On the other hand, a case may not be totally resolved by means of better communication or by getting more facts. That does not necessarily mean that the case presents a true ethical dilemma. Sometimes conflict can occur as a result of a lack of clarity about the particular demands of a single theoretical approach. For example, all parties to a case may agree that the recognition of a patient's autonomy is of primary importance, but may disagree on how best to achieve that recognition. Thus, all may agree that a patient's right to self-determination entails that he has the right to fully informed consent, but they may disagree about the amount and type of information that the patient needs in order to make a truly self-determined choice. Or, for another example, it may be agreed that a utilitarian approach is appropriate to making a particular decision, but there may be disagreement about the probable consequences of various possible decisions. Thus, all may agree that it is important for society to allocate money for dealing with cancer; but they may disagree about whether limited money is better used for prevention or for treatment.

These types of problems are also amenable to solutions. Finding solutions may depend on a careful analysis of available facts, as is probably the case with the specific issue about informed consent. The solution may have to wait until more general facts are gathered and analyzed, as is probably the case with the issue of the allocation of money for cancer care.

Finally, there are the real ethical dilemmas. It may be that all the relevant facts are in, but there is a fundamental disagreement about the most ethical resolution. At this point one must choose the theoretical framework that he believes is most intellectually justified. At present, the general trend in health care decisions involving individual patients is towards a recognition of autonomy theory as the guiding principle. (The author of this book is in agreement with that trend, and readers are cautioned that a bias towards that approach may be evident in this book.) That is not to say that the autonomy approach is the indisputably justified way.

I have heard students say that they prefer to be flexible in their approach to issues, and would prefer to make decisions on a case-by-case basis. If they mean that they intend to keep an open mind, all the better. However, they more often mean that they prefer to use utilitarian approaches at one time, autonomy approaches another time, and so forth. That approach will not work, for a number of reasons. First, they should examine the reasons they would choose one approach in some cases, and another approach in other cases. Presumably the decisions to change horses would not be haphazard. It may be that they have not really examined their underlying premises. If they are haphazard in their approach, they will not only be guilty of a charge of arbitrariness, but will find that there will be cases in which they will be caught in impasses of inconsistency.

Chapter 2

Informed Consent and the Right to Refuse Treatment

The increasing rejection of paternalistic approaches to medical care and the growing stress on patient autonomy are among factors creating many new legal obligations for physicians. Among these obligations are the requirements that cluster around the concept of "informed consent." More and more, physicians are being required to allow their patients the opportunity to accept or reject pro-offered procedures and also, in many cases, being required to give their patients information about the procedures.

CONSENT

The requirement that patients be allowed to refuse pro-offered procedures—the "consent" part of "informed consent"—is really an unsurprising implication of the acceptance of autonomy theory. The ability to choose is fundamental to the meaning of autonomy. To deny someone the opportunity to choose to accept or reject a medical treatment, even with the intent of benefiting him, is to deny his self-determination.

The courts have often interpreted a failure to get a patient's consent for an intrusive medical procedure as a battery. That is, the law sometimes treats the action of the physican in essentially the same way as it does a criminal attack or any "unwanted" or "offensive" touching of a person.[1] The early cases focused on the question of whether actual harm was caused to the patient as a result of the failure to get consent. That is, the approach to the cases was utilitarian. If the procedure was one that the patient, or "a reasonable person," probably would not have consented to if he had had the opportunity, and if the procedure resulted in harm to the patient, then the courts would find the physician guilty of battery. If there was no harm, then the courts would not find against the physician. That approach seems odd. In most cases the intent of the physician's "battery" is not malicious, as it is in ordinary criminal battery. Moreover, in most cases the physician is not engaging in an act that is reckless and likely to cause harm. For example, he is not casually throwing scalpels around the hospital cafeteria. Simply interpreting a failure to get consent as a battery doesn't capture the whole story.[2] Perhaps it is best to interpret the courts' approach in presenting such cases as instances of bodily harm as an evolving attempt to grapple with the implications of the rights related to autonomy. It seems more appropriate to think of a failure to get consent primarily as a violation of the patient's autonomy rights. Justice Cardozo's classic expression of the right to informed consent seems to reflect this interpretation:

> Every human being of adult years and sound mind has a right to determine what shall be done with his own body; and a surgeon who performs an operation without his patient's consent commits an assault for which he is liable in damages . . .[3]

In general the law seems to be coming around to that point of view. More often, now, a failure to get an informed consent is thought of as a violation of a patient's rights, whether or not the patient suffered injury as a result of the treatment. So, for example, in a recent case regarding the withdrawal of life-sustaining care, the judge wrote in regard to choice:

> Medical choices are private. . . . They are not to be decided by societal standards of reasonableness or normalcy. Rather, it is the patient's preferences—formed by his or her unique experiences—that should control.[4]

The philosophical justification for the requirement that physicians give an explanation of the proposed procedure—the "informed" part of "informed consent"—is a bit more complicated.

A *minimum* requirement for respect for patient autonomy might leave the physician simply with an obligation to allow the patient the opportunity to refuse an offered procedure. That would be a view of medical care that worked on the economic model of "let the buyer beware." Such a model gives the patient a choice whether or not to accept a procedure, but leaves him with the responsibility to seek out information about the risks and alternatives to the proposed treatment. In the language of rights, that model gives the patient a *negative* right or liberty to accept or refuse treatment, but not a *positive* right to be given information about the treatment.

The law, at least, does recognize such a minimal view of the physician's obligation in regard to procedures that could be considered to be commonly understood by the public or by the "average person," and which, as well, pose no substantial risk to the patient. So, for example, there would be no *legal* obligation to offer to fully explain the process, purpose, and risks of taking a blood pressure if the patient didn't ask for an explanation.[5] Nor would there be a legal obligation to present the patient with a form to sign verifying that the procedure of taking a blood pressure has been explained to him and that he has consented to it.[6] The procedure is common enough to be considered to be within the knowledge of the average person and, as well, the procedure usually poses little risk.

But the law, in assuming that certain procedures and their risks are common knowledge, is *not* implying that those procedures can be forced upon an unwilling patient. The implication is only that the average layperson is expected to know enough about the procedure and its benefits and risks to be able to make a choice to accept or reject it without a formal explanation.[7] Thus, although a physician doesn't have to get a written consent for the blood pressure taking, the patient nevertheless may have the right to refuse the procedure.[8] If the patient does not refuse, then there is an assumption that he has consented—that is, there is then an assumption of "implied" or "presumed" consent.

INFORMED

The law recognizes stronger obligations on the part of the physician in regard to certain classes of procedures, not only recognizing the patient's right to refuse them, but also recognizing an obligation of the physician to give the patient an explanation of the procedure, information about its risks, its benefits, and information about alternative treatments that may be as appropriate. The "risks" that must be described may not be limited to medical risks. For example, many hospitals require specific informed consent for HIV antibody tests. The rational is not that the test poses a significant physical risk, but that a positive test result, or even the fact that the patient has been tested for the antibody, may pose various social and economic risks to the patient if information were to become known to employers or insurance companies.

There may be a further legal requirement that there be written evidence of the explanation and consent, signed by the patient, the physician, and a witness.

All this, of course, is what we usually mean by "informed consent."

Roughly speaking, laws at present tend to require the full explanation and written consent when the procedure is complex and/or is attendant with substantial risk. The justifications that could be given for the duty to provide the patient with information are that (1) the layperson, no matter how "autonomous," cannot be expected to have enough information about complex medical procedures to enable him to make a truly "voluntary" choice; (2) the public will not permit the practice of medicine and the physician-patient relationship to be looked upon *simply* as the selling of an item or service by vendor to vendee; society will not permit a "caveat emptor" in the "selling" of medicine, and thus, through means of the social contract, makes disclosure a requirement; and (3) the belief that information given to the patient can enable him to aid in his own treatment.

In more fundamental philosophical terms, the justification for the duty to inform could be found in autonomy theory and in social contract theory.

"Informed consent" is among the most confusing of the ethical concepts. Part of the reason for the confusion is that the concept itself and the legal requirements tied to the concept are still evolving. The scope of the legally recognized right of patients to informed consent, like many newly recognized and evolving rights, still has unclear borders.

To complicate matters, issues relating to informed consent are tightly tied to many of the other ethical issues in medicine. In fact, it may be possible to see issues such as truth-telling, and the right to refuse treatment, as problems subsumable under the general issue of informed consent. In that regard, the reader should refer to the appropriate sections in this book.

As we have said, early legal requirements in regard to informed consent for treatment focused only on procedures that are thought to pose a substantial risk to the patient. As a result of that legal history, many physicians believe that they are required to "get" an informed consent only for those procedures that may pose a serious physical risk. But that belief is a mistaken oversimplification of what may be morally, if not legally, entailed by the "right to informed consent."

It is difficult to avoid dwelling on legal aspects when speaking of the issue of informed consent, but we will try to do just that from here on. Our discussion will center on the ethics of informed consent and we will keep discussion of the law on the edge of our discussion.

There are three aspects of informed consent that are of immediate importance to physicians in training: (1) those aspects relating to treatment; (2) those related to medical education; and (3) those related to research. In many instances, issues relating to all three will come up in one case.

We will focus on those issues relating to treatment

and education. While research issues are extremely important, they usually do not usually play a primary role in medical education; so we will reserve discussion of them for the last chapter.

IN REGARD TO TREATMENT

Competent Patients

There is a sense in which humane medical treatment already includes a moral requirement that patients should be at least somewhat informed about intended medical procedures. That requirement comes from the recognition that medical treatment involves interactions between persons, rather than simply the actions of a person (the physician) upon a thing (the patient). For example, it would be both a lack of respect and a lack of humanity to walk into a hospital room and take a blood pressure of a conscious patient without speaking to him at all. Minimum considerations of decency and respect would require the physician or student to at least say *something* to the patient about his intentions rather than just walking over to the patient, and grabbing his arm. Speaking to the patient acknowledges that you are dealing with a person rather than simply with a disease or organ.

Subtle differences between morality and law emerge here. If the patient makes no protest to a physician who walks in and takes his blood pressure without a word, then presumably the patient has given his implied consent to the blood pressure taking and thus, likely, no law has been broken.

Many patients are quite compliant to such forms of treatment. For example, the medical student who is taken on rounds may find himself taking blood pressures (among other things) of patients to whom he has not been introduced in even a perfunctory way, without the patient complaining. He may find that the patient is seldom or never spoken *to* directly, but instead spoken *about*, and then only as exemplar of an illness.

Sometimes the student may find that he is introduced to the patient by name, but that the dynamics of the interaction quickly shift over to a discussion between preceptor and students, with the patient being spoken over and about, but not with. A well-meaning attempt to respect the personhood of the patient can easily deteriorate in the rush to "get clinical material" and with the desire to show off one's medical knowledge to preceptors.

The law might permit such treatment if the patient makes no protest, but regardless of the law, it might be said that such an "interaction" goes contrary to those moral principles that require respect for persons.

However, there are issues about fulfilling even the minimum requirement of saying "something" to the patient, and it is easy to think of a number of real ethical issues that arise even in regard to something as simple as taking a blood pressure:

1. *What* should the physician (or student) say to the patient upon entering the room?
2. Does the physician have an obligation to explain to the patient precisely and fully what he will do with the information he gets from the blood pressure taking?
3. Does he have an obligation to give the patient information about blood circulation, the workings of sphygmomanometers, and so forth?
4. Does the patient have the right to refuse the tak-

ing of the blood pressure if the physician feels the information is necessary to the care of the patient?
5. If the patient acquiesces to taking the blood pressure without asking the reason, does the physician have any obligation to volunteer an explanation to the patient?
6. Does he have an obligation to tell the patient if the patient specifically says he doesn't want to know what the purposes are?
7. Suppose he knows that the patient is likely to present with a higher pressure due to situational anxiety: Should he distract the patient without telling the patient that he is distracting him and why is he distracting him?
8. If he is a medical student, does he have an obligation to correct the patient who says "Here is my arm, 'Doctor'"?
9. If the patient is incompetent, must he get a surrogate's consent for a procedure like taking blood pressure?

These are the sorts of questions that arise in regard to consent for procedures, whether or not the procedures have great probable risk.

When the procedure is more serious, or the patient is incompetent, or we are dealing with a research subject, there are further complications of which we will speak later.

Under an autonomy approach, the patient who *requests* an explanation of a procedure, no matter how routine or safe the procedure is, has the right to a full explanation of the purpose of the procedure.

This is expressed in the American Hospital Association's *Bill of Patient's Rights*:

> 6. [You have the right to] Receive from your physician information necessary to give informed consent prior to the start of any [non-emergency] procedure and/or treatment . . .[9]

The statement does not limit the patient's right to informed consent to serious or complicated procedures.

The physician may voice a number of objections to such a demand: "How can I be expected to explain something to a layman in five minutes that took *me* hours or weeks of hard and complex work to learn? Moreover, I don't have time to explain *every* little procedure like pulse taking or pressure taking to patients[10]—it takes valuable time that I need to treat those very patients.

"And, in any event, whether or not the patient knows the purpose of minor procedures will make no difference in his care and his ignorance will place him in no great risk. After all, I am trying to help him. Besides, patients are often irrational—a patient is likely to get panicky about a statistically insignificant risk in surgery, or likely to develop side effects of a medication if he hears about them.

"Moreover, he will quickly forget what he is told anyway."

In spite of those real or imagined difficulties, the requirement is that the physician make an attempt to explain his proposed treatment to the patient in a way that the patient can understand. Many physicians believe that the obtaining of informed consent is simply an administrative or legal formality and a nuisance. They may look upon "getting" the consent as they look upon getting any bureaucracy-engendered form, and believe that they have fulfilled their legal obligations once they have obtained the patient's signature on the hospital's form. Thus the still common practice of a resident or attending sending a medical student in to "get a signature." Not atypical is a "getting" of consent such as the following: A Hispanic female, twenty-five years old, has been admitted to the hospital with a fibroid cyst. She speaks almost no English.

The resident approaches her and tells her in English that she will "need an operation," but not to worry. He leaves. Two hours later, the resident asks a medical student to bring her the informed consent form for a signature. He approaches her and, in English, asks her to sign the form. She signs.

Has the woman given consent? She has signed the form, but the fact that a patient signs a consent form does not morally or even legally prove that the patient has given consent. A simple complete explanation in English to a patient who has little or no command of the language cannot count as informing the patient.

Nor does the fact the information has been given in the patient's native language necessarily satisfy the demand to inform. The physician should be aware that his training has immersed him in jargon. Much medical terminology has become second nature to him, and he may unthinkingly assume that patients understand such terminology. A patient, not wanting to appear stupid, may pretend to understand. In one instance, for example, a patient was given a medical history questionnaire. Among the questions was "S.O.B.?" Most laypersons do not take those to be the initials for "shortness of breath."

A useful way of verifying a patient's understanding is to ask the patient to explain back the information in his own words. That approach also has the advantage of giving a physician a general idea about patients' levels of understanding, as well as giving him feedback about his own ability to give understandable explanations.

Whenever possible the physician should give the patient time to think about the pro-offered procedure, and an opportunity to formulate questions about what he has been told. Patients are often fearful to an extent that interferes with their ability to digest information. It is undeniable that the information needed to make an informed choice may be a burden for the patient. The seriously ill patient, in pain and anxious, may be faced with a number of complex possible treatment options. A "cooling-off" period can give them time to think more clearly about what has been explained to them. If family and friends are available it may be helpful to encourage the patient to speak with them, preferably in the presence of the physician. If the claim that patients quickly forget information given to them is true, then that may simply be another reason to give them a chance to think about the procedure, time to ask questions, and time to discuss the options.

It is also true that a patient facing surgery may perceive a "statistically insignificant" risk as very significant. Yet here too a recognition of a patient's "sphere of autonomy" does not permit the physician to second-guess a patient's reaction to facts, nor does it permit him to bypass a decision the patient may make (or might have made had he known the facts) that goes contrary to the physician's own judgment.

In regard to explaining the possible side effects of medication, the utilitarian-consequentialist argument that side effects may be psychologically induced if the patient is warned about them must be balanced by the possible harm that nondisclosure poses if the patient were to develop side effects without knowing their source or significance. Moreover, such a consequentialist claim requires hard confirmatory data rather than unsubstantiated surmise about the probability of side effects being induced by disclosure.

Of course the autonomy argument starts with an assumption that the patient has the right to know the possible side effects of medication.

PLACEBOS IN TREATMENT

Informed consent in regard to medication is sometimes seen to give rise to a serious dilemma when the use of placebos seems a viable option in pain control. Of course, truly informed consent is really not possible when placebos are used. To avoid directly lying, some physicians recommend that the patient be told that he is being given "'something' that will relieve his pain." While that is not an outright lie, it still seems inconsistent with a patient's right to fully informed consent. The argument given for withholding truth or lying about placebos is that the intention is to benefit the patient by avoiding possible side effects of analgesics and also to avoid the possibility of the patient becoming addicted to narcotics. Given that there is a strong probability that the placebo will work, those are powerful arguments. As far as autonomy approaches are concerned, it is doubtful that either argument would outweigh the patient's rights to informed consent and to the truth. The claim that informed consent is not necessary because there is little risk in using placebos is a weak argument. Given that the analgesic effect of placebos is less probable than that of "real" drugs, the danger of using them may be less, but the risk in regard to pain alleviation is greater. The possible risks of telling the truth about the placebo and thus having to resort to "real" analgesics probably are neither serious enough nor imminent enough to warrant either direct lying or intentionally withholding the truth.

The use of placebos is theoretically consistent with utilitarian approaches. However, the full consequences of the use of placebos would have to be weighed very carefully. For example, what will happen in the future if the patient were to approach another physician with a request for "the same painkiller that Dr. Placebo gave me." One would also have to weigh the effects of lying upon the rest of the treatment team. Furthermore, we would have to consider the general effect upon patient care and trust if the hospital's use of placebos became widely known. For example, there are reverse placebo effects, and a real analgesic may not be as effective if patients come to believe that they are really receiving placebos.

WAIVING INFORMED CONSENT

Suppose a patient is not interested in the details of a procedure? May a patient waive the exercise of his right to informed consent?

This is not an uncommon situation. Some patients, either because of anxiety about their illness, or because of a fear that they are incapable of understanding medical information, or because they trust the physician, or simply because they are not interested in knowing, may simply say "Don't bother me with the details, Doc, just do what has to be done."

Those physicians who would prefer to avoid the difficulties of obtaining a fully informed consent may welcome these very compliant patients. Of course, things are not that simple. Autonomy theorists would argue that a patient has an obligation not to waive his right to informed consent. A patient's dignity and respect for his own autonomy should compel him to remain an active participant in decisions that concern him. Just as a person has an obligation to respect the autonomy of others and thus avoid infantilizing them, so should he respect his own autonomy and refuse to allow himself to be infantilized. However, given that the patient's "irresponsible" behavior doesn't cause immediate harm to others, it is unlikely that the autonomy theorist could argue that the physician should force the information upon the patient.

There are utilitarian arguments to support a duty to encourage, if not require, patients to be active participants in their own care. The patient who has information about his illness will be able to be on the lookout for complications. He also will be more able to cooperate usefully in any regimen necessary to his own recovery. After all, true "compliance" not only means consenting and passively submitting to a procedure but also doing what is necessary to restore health after the procedure is done. In certain instances, such as mental health care or physical rehabilitation, the active participation of the patient in his treatment may be part and parcel of recovery.

Nevertheless, it is problematic whether a patient should be forced to be an active participant in his care. But, at the least, patients who avoid being active participants in their own care should be encouraged to take a more active role. If the patient asks "What should I do, Doctor?" there may be nothing wrong with the physician saying "If I were you, I would . . ." However, in such cases, the physician should not be glib in his reply, but should try to place himself in the patient's situation before replying, and should add, "But the final decision is yours."

Coercion

> George S. is an alcoholic arrested for disorderly conduct. The judge gives him the choice of 30 days in jail or joining an alcoholic rehabilitation clinic at a local hospital. He agrees to go to the program, but sometimes is not compliant with the requirements of the program. The psychiatrist in charge tells him that unless he is more cooperative, the psychiatrist will see that he is sent to jail.

When we speak of consent, we always have in mind *voluntary* consent. If a patient agrees to a procedure as a result of duress or coercion, we really cannot say that the patient has given consent.

Coercion can range from outright threats to psychological abandonment. It is sometimes difficult to determine whether or not a situation is in fact coercive. For example, it might be said that the physician who expresses anger at a patient who refuses or is reluctant to accept the physician's recommendation is coercing the patient. Contrary to that, it could be claimed that an autonomous patient should be capable of being assertive enough not to let the physician's attitude sway him.

If, in trying to obtain a patient's consent, the physician presents a warning about consequences to the patient that the physician believes is unreal, then the physician is using coercion. Thus, a physician who knowingly and falsely predicts dire consequences if a patient doesn't take his advice is being coercive. On the other hand, the physician who really believes that a patient who refuses a procedure will probably die cannot be said to be coercing the patient if he gives voice to that belief.

Cases such as that of the alcoholic presented above

are more difficult to analyze. It is not unusual for the courts to offer defendants convicted of crimes involving the use of alcohol or narcotics the "choice" of diversion to treatment rather than imprisonment. On the one hand, such a choice may seem not only to be reasonable, but really doing a favor to the defendant. On the other hand, we may feel uneasy about the voluntariness of such "choices." Perhaps one source of our unease is the underlying belief that sending the defendant to prison presumes that he is responsible for a crime, while sending the defendant to treatment implies that the defendant's illness was a cause of his criminal behavior, and therefore he was not totally responsible for his actions. Thus the implication of the situation is that George S. is being threatened with punishment unless he cooperates in the treatment of his "medical" disorder.

Difficult too is the question of whether it is coercive for a hospital or physician to say to a patient "We believe that X treatment is the only appropriate treatment for you, if you don't accept that we will have to discharge you." We will discuss that issue in the section "The Right to Refuse Treatment."

SEEMING EXCEPTIONS TO THE REQUIREMENT OF INFORMED CONSENT

Emergency Procedures

Acute emergency situations are probably the only *real* exceptions to the requirement for informed consent. The other exceptions, which will be discussed, may really be *apparent* exceptions.

If the patient is in a life-or-death crisis and there is no time to obtain his fully informed consent, then a "common sense" judgment is made that physicians have a presumed consent to do what is necessary to save the patient's life. The rationale is the belief that *most* persons probably would want to be saved in such situations.

There may be compelling reasons to treat such patients even if they refuse treatment in the emergency room. If the patient has not received full information about an emergency procedure that is believed necessary to save his life, some might argue that his inability to give consent to the procedure entails that he hasn't the ability to refuse the procedure either. Thus, for example, there may be compelling reasons to give blood to the emergency room patient who protests that he is a Jehovah's Witness if, in fact, there is no time to explain to him the full implications of his wish to refuse blood.

> A 60-year-old man is brought to the emergency room unconscious. He is accompanied by his wife. The resident on call is somewhat familiar with the patient and knows that he is in the end stages of a terminal disease. He begins to suffer cardiac arrest. As an attempt begins to resuscitate him, his wife produces a note with the patient's signature which says that he does not want to be resuscitated if he were to suffer cardiac arrest. She asks that his wishes be honored.

While there appears to be some evidence that the patient does not want to be resuscitated, that evidence (as described) does not seem conclusive. For example, there is not enough information about the patient's condition or his chances for recovery. Concomitantly, there is no evidence that the patient knows the consequences of either accepting or refusing resuscitation. It is probable that in such cases it would be unwise to go along with an unconfirmed and uninvestigated "living will."

The situation in an emergency room, or when an unexpected acute episode occurs in the hospital, must be contrasted with a situation in which a now incompetent and seriously ill patient has left clear, documented, and explicit instructions about his care before becoming incompetent. For a more detailed discussion see Chapter 4, "Euthanasia."

Incompetent Patients

Issues having to do with informed consent for incompetent patients are among the most pressing problems for medicine.

Here too, trends towards an acceptance of autonomy approaches are having great effects on the ways in which decisions are made for those who are incapable of making their own decisions.

The two trends that are having the most impact are (1) the tying of competency to specific contexts and (2) the attempts to form "substituted judgments" for incompetent patients.

"Competency" and "incompetency" are really legal rather than medical or psychiatric terms. Properly, we should be speaking of "decisional capacity or capability" rather than of "competency." Most states allow hospitals to "slur" the distinction, and do not ordinarily require a court hearing to substantiate that a patient evaluated by psychiatrists as "mentally or decisionally incapacitated" is legally incompetent. However, a patient would certainly have the right to request a judicial determination if he believed that a determination of incapacity made by a physician or psychiatrist was unjust.

Competency Related to Context

> R. is an involuntary patient in the psychiatric ward of a municipal hospital. A court has awarded power of attorney for his financial affairs to his wife. He was admitted to the hospital because of suicidal ideation stemming from depression. Treatment with imipramine has had limited success. A reevaluation of his case has led the attending psychiatrist to believe that lithium treatment would be a better modality than the imipramine. He writes an order for lithium. R., noticing the change in medication, expresses some reservations about taking the new medication. He asks the medical student assigned to him about the medication change.

In the past, patients who were involuntarily committed to mental institutions lost many of their rights while committed. There was an assumption that anyone who was psychotic enough to be committed was incapable of making any reasonable judgments about any matter.

That situation has changed,[11] and a growing number of court decisions support a belief that a diagnosis of psychosis, or status as an involuntarily committed patient, are not equivalent to judgments of global incompetence. The belief is that a patient can be incompetent in some areas of functioning yet competent in other areas. Since the patient must be considered competent until shown otherwise, *the burden of proof is on the physician to show incompetence;* not upon *the patient to prove his competence.* This has meant not only a recognition of a right of mental patients to refuse treatment but, along with this, a recognition of their right to informed consent.

Thus, in our case example, neither the fact that the patient is an involuntary commitment nor the fact that he has been determined to be incompetent to handle his finances are sufficient reasons to automatically make him incapable of decisions regarding his treatment. Given that lithium has serious side effects, he would have a prima facie right to be told of those side effects, the expected benefits of the treatment, alternatives to the lithium, and the probable consequences of alternative courses of action. He then would have the prima facie right to accept or refuse the lithium.

If there is a belief that he is not capable of making a decision in regard to his treatment, then his capacity to make a decision would have to be evaluated. Essentially, that evaluation would be made on the basis of the patient's ability to understand the content of the informed consent in regard to treatment.[12] If, for example, the patient shows an understanding of the benefits of lithium treatment, the risks of accepting and not accepting the treatment, he would likely have to be considered capable of making a choice about the treatment.

> T. is an involuntarily committed psychiatric patient. There is reason to believe he has a thyroid cancer, and a recommendation for biopsy is made. T. is approached for consent. He replies: "Yes, it's all right to do the biopsy. I spoke to my friend who said that only Martians can treat me. I know you are really Martians, so it's OK."

Problems may arise with compliant psychiatric patients too. The physician may encounter psychiatric patients who agree to treatment, but appear to be incapable of making a decision. In such cases it may seem easy to simply accept the patient's consent, but if in fact the patient was incapable of giving consent he would have to be treated as an incompetent patient.

Incompetency in Regard to Treatment

Temporary Incompetency. There are times at which a patient facing a serious procedure is in a temporary state of mental incapacity. The most obvious example would be a patient who is asleep. More common and cogent examples include patients who are temporarily unconscious, or situationally agitated, or under the influence of medication that seriously affects their thought processes.

In these cases, given that there is no medical emergency, and given that the patient could be brought back to competency without causing him serious harm, there is an obligation to bring the patient back to competency and then attempt to obtain his consent.

> W. is admitted to the hospital with abdominal pain. He has a history of gallstones. Preliminary examination reveals a gallstone flare-up, and surgery is recommended to remove the gallbladder. W. is approached by a surgical resident and informed of the diagnosis, the recommendations, the risks, and the probable consequences of refusing surgery. He tells W. that while antibiotics will probably control this acute episode, he will continue to have problems unless the gallbladder is removed. W., although clearly in pain, says that he understands, that he has thought carefully about the alternatives, and says that he does want antibiotics, but does not want surgery. The resident leaves to speak to the Chief of Surgery. The Chief of Surgery says that he has spoken to W.'s wife, who wants her husband to have surgery. He tells the resident to wait. That evening, W. becomes feverish and semi-coherent. The resident on call believes that the patient could be treated conservatively and brought back to competency by morning. Nevertheless, a psychiatric consult is called, and W. is evaluated as incapable of making treatment decisions. W.'s wife is called and she signs consent for the surgery.

The practice of waiting for a competent "uncooperative" patient to become incompetent is not uncommon. It should be clear from the analysis of informed consent given here that this practice is, at the least, questionable. If the patient while competent had full information about alternatives and was capable of making a choice, then, given that nothing unpredictable occurs, he has the right that his wishes be followed even if he *becomes* incompetent.

Surrogate Decisions. When a patient is truly incapable of making a decision in a nonemergency situation, and he has not made his wishes known prior to becoming incompetent, a surrogate must be designated to make the decision for the patient.

Autonomy approaches are having important effects on the concept of surrogacy. In the past, a surrogate for adult patients was usually perceived as someone designated to decide what is in the "best interests" of the incompetent patient. Essentially, that is a paternalistic approach to surrogacy. The present trend perceives the surrogate of an adult patient as someone designated to give voice to what *the patient would have wished* were he competent. The surrogate is to make what is called a "substituted judgment." If that is not possible, then the surrogate decides according to the "best interests" principle.

> G. and R. are brothers. G., a Jehovah's Witness, is now incompetent and facing surgery which will require whole blood transfusions. R. has been designated a surrogate for G. R. has always looked upon G.'s religious beliefs as unacceptable. However, he knows that his brother was always true to his own religious beliefs and on a number of occasions had expressed his unwillingness to accept whole blood.

While R. may fully believe that it is in his brother's

best interests to receive blood, his role as a surrogate who is charged with the obligation to make a substituted judgment would require him to refuse to sign the consent for transfusions.

There are very difficult issues inherent in both the "substituted judgment" and "best interests" approach.

Many patients do not explicitly make their wishes known about treatment before becoming incompetent. There has been a great deal of movement towards encouraging the use of living wills and the assigning of a "durable power of attorney for medical treatment." In the latter instances, patients while competent can name someone to be empowered to make decisions for them.

If a patient hasn't designated a surrogate, then a surrogate must be chosen for him. Hospitals often have a policy of accepting the decision of a patient's next of kin. In line with what we have just said about the role of a surrogate, it should be apparent that such policies are neither automatically legal nor necessarily morally appropriate. The next of kin may not be an appropriate surrogate for a number of reasons.

The next of kin may not know the patient well. For example, the next of kin of elderly patients may be geographically and emotionally distant from the patient.

Furthermore, the next of kin may have agendas that are inconsistent with his role as surrogate. He may be feeling guilt or anger about his relative's illness. He may stand to gain financially from the death of his relative.

There may be other persons who know the patient well and are able to testify about the wishes of the patient. These may be so-called "significant others" such as lovers or friends, or clergy, or even neighbors. With patients who have been in hospital for long periods, a nurse or social worker may be the person most capable of forming a substituted judgment.

Of course, such significant others may have the same underlying agendas as a next of kin.

Adding to the difficulties, even a dispassoniate person who knows the patient well may have a hard time making a substituted judgment. People seldom think clearly and fully about medical decisions before they face a serious illness and thus seldom voice wishes about treatment in the event of incapacity. If they do express wishes, the wishes seldom take into account the contingencies that probably will come up in the actual situation. It is hard to project one's own wishes towards such situations, and even harder to project what someone else's wishes would be.

A physician who is attempting to get a consent for a procedure for an incompetent patient should be aware of these difficulties before simply allowing a next of kin to sign the consent for a procedure. If there are complex treatment alternatives and/or the procedures are serious, the physician may be forced to spend some time trying to find an appropriate surrogate. If his hospital's policy is to accept a next of kin's decision, the physician may have a moral duty to educate the next of kin about his role as a surrogate. If he suspects that the next of kin is an inappropriate surrogate then he may have an obligation to hold off getting a consent signature until he confers with the hospital lawyer or administrator or ethics committee.

We should also stress the importance of coordinating care with the whole core treatment team. Nurses, clergy, social workers may have information about the patient and the patient's family that could be crucial to making an appropriate decision.

Still, there remains an issue about who has the obligation to find a surrogate. On the one hand, the courts would be overloaded if they had to adjudicate every case that required a surrogate; on the other hand, it is not clear that the designation of a surrogate should be part of the duties of physicians.

There are special instances in which forming a substituted judgment based on the incompetent patient's former wishes and character would be difficult, if not impossible.

The two types of cases that come to mind are children and very retarded persons.

The issue of infants is discussed in Chapter 4, "Euthanasia." To adumbrate what is said there, a determination would have to be based on a "best interests" principle. The difficulties with applying that standard to defective newborns is discussed in that chapter.

Older children present a different issue. The law is utilitarian in attributing responsibility to children. Although there may be individual minors who are able to take on various responsibilities quite early, for simplicity's sake the law generalizes and chooses arbitrary ages for the assigning of "majority." Moreover, the legal age of majority varies for different contexts—e.g., drinking, marriage without parental consent, driving. Below those ages, minors are assumed to be incompetent. In the case of medical decisions, the law allows parents to make decisions for children, working on the assumption that parents will act in their children's "best interests." However, the state always reserves the right to override a parental decision that it believes to be against the best interests of the minor. The usual medical examples cited are instances in which the state gives a guardian temporary custody of a child of Jehovah's Witnesses to insure that the child will get needed whole blood.

Moreover, the law may allow minor children to prove that they are "emancipated" (not dependent on their parents), and hence able to make their own decisions regarding certain matters, including health care.

The issue of a minor's ability to give consent for medical procedures has recently come up in regard to consent for abortions. The Reagan Administration, giving a utilitarian argument that the state has an interest in promoting "family values," supported legislation requiring that a minor female who seeks an abortion show the physician proof that her parents have been notified of her decision. The Administration also supported legislation requiring parental consent for minors who seek abortions.

The arguments against such legislation are given on both utilitarian and autonomy grounds.[13] One utilitarian counterargument is that such policies will have the unhappy effect of forcing the birth of many unwanted children. Autonomy counterarguments claim that, first, the state should not be vested with the duty of forcibly promoting "family values"; second, that the family values being promoted conform to a particular bias about the desirable nature of the family; and third, though not least important, that a minor pregnant female should be considered competent to give consent to abortion.

At this writing, the Supreme Court is in the process of examining the constitutionality of such legislation.

The physician treating a minor should be aware that there may be moral reasons either to allow the minor to make a decision or at least to take part in making a decision regarding his medical care. If he has evidence that the minor is capable of understanding what is at stake in his treatment, then it may be the case that the child should have something to say about his treatment. At a minimum, that would entail that the physician give the minor information about his condition and proposed treatment. In instances in which there is conflict between parents and a child who appears to be knowledgeable about his condition and exhibits emotional and intellectual maturity, the child's wishes may have to be taken into account. This may place a burden upon the physician to try to resolve the conflict between parents and child and, if that doesn't work, perhaps refer the case for consideration by the patient advocate, the hospital ethics committee, or the courts.

The issue of consent for very retarded persons is another matter. First, the partitioning of the concept of competency would probably require that a specific determination be made that the retarded patient is not capable of making a decision about the particular treatment in question. That is, the fact that he is retarded or institutionalized would not automatically mean he is incompetent to make decisions about his medical treatment. The retarded patient's ability to give consent may depend both on his degree of retardation and on the complexity of the medical decision. For a patient who is too retarded to make a particular treatment decision, a "substituted judgment" based on prior wishes is likely to be impossible because he probably never had a capacity to formulate any comprehensive desires concerning treatment.

Similarly, the suggestion that a decision be made on the basis of what a "reasonable person" would choose may also be inappropriate. The retarded patient's ability to cope with pain or to cooperate in his rehabilitation may be a function of his very ability to understand what is involved in his disease and treatment. Thus, the subjective effects and consequences of treatment upon a very retarded person—factors that normal patients will likely comprehend, and are important to their decision—may be difficult to determine from the "normal" person's point of view. These difficulties of determining "what a reasonable person would decide were he retarded" are self-evident.

The more feasible approach might seem to be to make decisions based on the "best interests" of the patient. Yet, that approach faces similar difficulties, for the "best interests" of the retarded patient will inevitably be decided from the standpoint of the normal person who is chosen to be surrogate.

The most famous recent court case confronting this problem was the case of Saikewicz.[14] Saikewicz was a profoundly retarded adult living in an institution. He was diagnosed with a leukemia. The belief was that chemotherapy could prolong his life, but not cure him. A court-appointed guardian claimed that treatment would be against Saikewicz's best interests. The guardian argued that because Saikewicz was incapable of comprehending the reason for treatment, he would suffer pain and fear which would outweigh the limited benefit of the treatment. The court believed that it was possible to use the substituted judgment criterion in this case. It decided that if Saikewicz had been able to make a decision, given his circumstances, he would have decided against treatment.

Given the difficulties we have already described, one might easily question the court's decision in this cases.

Therapeutic Exceptions

Sometimes a claim is made that a competent patient's right to informed consent may be overridden if it is judged that discussing the proposed treatment with the patient will cause him serious harm. This so-called "therapeutic exception" is a sticky issue. Questions could be raised about what constitutes "serious harm," as well as questions about the predictability of serious harm. Some "harm" might be considered quite natural. For example, it is likely that any normal person confronted with the possibility of major surgery would get anxious, and that any person confronted with a poor prognosis would get depressed. The therapeutic exception may open the possibility of misuse. For example, it might serve as a rationalization for physicians who are uncomfortable discussing difficult choices with their parents.

The Right to Refuse Treatment

Implied in the requirement of patient consent is the right of the patient to refuse to give consent—that is, the patient's right to refuse treatment. We have touched on many of the issues pertaining to the right to refuse treatment already, and some of the issues we reserve for other chapters (Chapters 4 and 5). As we have said, there has been a growing acceptance that a competent patient has the right to refuse any treatment. That includes life-saving treatment, and treatments that the physician or even a preponderance of society might accept.

However, there are questions that could be asked about the *consequences* of refusing treatment. Can a physician withdraw from the care of a patient who refuses to agree to the physician's recommendations for treatment? Can the hospital discharge such a patient?

On the face of it, a strict autonomy view implies that the patient has the right to refuse treatment, but also seems to imply that the hospital or physician have no obligation to stray from what they believe to be the best treatment. Thus, the total implication is that the physician can refuse to treat the patient and the hospital may discharge the patient without violating the patient's rights.

However, the issue is far more complicated than it may first appear. First, it might be claimed that the approach just described is still coercive and is not, in fact, in accord with a recognition of patient autonomy. The reasoning is that the patient is seldom in a position to find other options for the treatment he wishes. The real availability of options may depend on the number of physicians and hospitals available in the area and, as well, may depend upon the ability of the patient to make use of other facilities. So, for example, an ambulatory patient rejecting a particular suggested elective procedure may have less of a right to stay in the hospital than would a seriously ill bedridden patient. The courts have not reached a consensus on this issue; in some cases they have forced hospitals to treat a patient on his terms, in others they have directed that another facility be found that would treat the patient on his terms, and in still others they appear to be upholding the right of the hospital to discharge such patients. The general movement seems to be towards allowing seriously ill patients to have it both ways—allowing them to reject treatment and forcing hospitals to keep the patients in the hospital.

EDUCATION ISSUES

Amy P. is a third-year medical student. Having entered medical school after another career, she is older than most other members of her class. Sent in to examine a patient, she is greeted by the patient: "Am I glad to see a real doctor, I don't want any more of those medical students poking at me."

There are a number of ethical problems that are unique to the education of physicians. Much of medical education is based on the model of "sink or swim" and medical students are often thrown into very deep water. Every learning situation is perceived as a test which will be graded by peers, by preceptors, by patients, and by the student himself. As we said before, the pressures to learn, to show off one's knowledge, and to get clinical experience are not always conducive to ethical medicine.

Amy P. seems to be faced with a dilemma. If she admits to the fact that she is not yet entitled to be called "Doctor," the patient will refuse the examination and she, in turn, will lose a learning experience. Not only will she lose a learning experience, but she may lose face with her fellow medical students and, furthermore, may run into trouble with her supervising resident.

These types of problems continue through residency. The often cited instance is that of "phantom" or "ghost" surgery. Here a patient consents to surgery to be performed by a particular physician but once under anesthetic he is operated on by a surgical resident.

Other related issues include the patient who objects to the flurry of students and residents circling around him during rounds.

There are many rationales given for such treatment. These include:

1. When the patient signs a consent form, it often states that the patient gives permission to "the physician or someone he designates to do treatment." The patient ought to have read the form.
2. The function of rounds is really to help the patient—the more students and residents see him, the more input there is about his case.
3. Medicine can only be taught in a "hands-on"

way; if patients were told the truth, they would refuse to be treated by students, and students would never learn medicine.
4. Patients who enter a teaching hospital should be aware that they may be treated by students. If they don't like that, they can enter a non-teaching hospital.
5. Lying or withholding the truth about one's status is innocuous. No one is sent in to treat the patient who isn't competent to do so, or who isn't being supervised by someone with proper credentials.

It should be pointed out that patients vary in their knowledge of distinctions among health care providers. While many patients have a fairly clear idea of the distinction between the work of nurses and the work of physicians, few probably have any knowledge of the differences between first-, second-, and third-year medical students, between residents and attendings, or even have knowledge of the distinctions among the subspecialities of medicine. The reader should think back to his ideas about these distinctions before he entered medical school. Even if he knows the distinctions now, the probability is that he didn't know them even as an educated undergraduate. Moreover, distinctions have become blurred in actual practice; for example, it is sometimes difficult even for hospital staff to tell the differences in the duties of nurse practitioners or physician's assistants and physicians.

The patient's concept of physicians may, in fact, be simply divided into two groups—"medical students" and "doctors." He may similarly divide their abilities in two ways: you are a medical student until you can do medicine, at which point they make you a doctor.

With that in mind, and given that the clinical clerk has been asked to do something within his abilities and duties,[15] a combination of the truth and an explanation may do the trick. For example, if Amy P. were to explain to the patient that she is a medical student but that she is at a stage where she has been trained to do the particular procedure, the patient might agree.

Of course the patient still may insist that he not be treated by a medical student, in which case, he may be a "clinical experience" that is lost to Amy and her colleagues.[16]

It might be claimed that the patient has an obligation to allow himself to be treated by students and to allow himself to be the subject of medical rounds. As stated above, the usual basis given for this argument is that the patient has voluntarily entered a teaching hospital and therefore has implicitly agreed to be "used" for teaching. But this argument is unrealistic in most cases. The real choice of a hospital may not be available to most patients. For example, there may be municipal or state statutes that limit the number of hospital beds in an area and that, in turn, may limit the options for patients. More specifically, those dependent on public means for paying for health care may be limited to being treated in specific hospitals. There also may be state or federal statutes that place a limit on the number of hospitals that can do a certain procedure (heart transplants, for example). Here, as in the right-to-refuse-treatment cases discussed above, the general thinking is that it would be too coercive to obligate patients in teaching hospitals to submit to situations in which the primary function is teaching rather than treatment.

And while an argument may be presented that rounds with medical students may be for the benefit of the patient, it is clear that the primary function of such rounds is to educate medical students. That is not to deny that a medical student can contribute in an important way to the care of the patient. In general, there has been more of a recognition of a right of patients to refuse to "submit" to teaching rounds.

Issues like those relating to "phantom surgery" by residents are somewhat different. There is an acceptance on the part of society, supported by licensing and certification laws, that residents are competent to perform many procedures, if more or less supervised. However, that does not imply that a patient can be misled into believing that a chief surgeon is going to do a procedure when, in fact, his resident is going to do most of the work. There are real as well as rationalized problems in defining "most of the work." Certainly it is a rationalization to say that the surgeon is doing the procedure if he is only standing by while the resident opens, operates, and closes the patient. On the other hand, there are many cases in which the surgery is, in fact, a joint effort.

The American Medical Association expressly forbids ghost surgery.[17]

Here too there is a strong argument against disclosure. The claim is that no one would willingly consent to have surgery done by a resident, and since it is important to train residents and impossible to burden attendings with all the surgery that has to be done, it is better to withhold the truth from patients. Moreover, as we said above, the consent form notifies the patient that a resident may be performing the surgery.

In regard to the consent form, it could be argued that such forms may be misleading. Even when time is spent going over the form with the patient, the section that mentions "Dr. X or someone he designates" is often skipped or slurred over in the explanation. We are faced here once again with the issue of "let the buyer beware"; can we expect the anxious patient facing surgery to be autonomous enough to spot that line in the form and, if he wishes, question its meaning?

If the patient himself is paying for surgery, then there is even a stronger argument to disclose. In spite of what is sometimes said, such patients are paying a high fee for a particular surgeon to *do* a procedure, not simply paying to have him *supervise* a procedure.[18]

There is little doubt that the general practice of concealing the fact that residents are doing procedures leads to widespread patient mistrust. The public may very logically conclude that if there were nothing suspect about residents doing surgery, there would be no reason to conceal the fact that they do. That in turn easily leads to patient suspicion about the qualifications of residents, and that, again, may easily lead to further concealment.

Given that residents are qualified to do procedures, and putting aside the large issue of two-tiered health care for rich and poor, there may be nothing wrong with an open system that both admits to the use of residents and requires that patients accept treatment by residents. That, in turn, may require disclosure of a type that is already beginning to occur on a large level. We are seeing a movement toward requiring hospitals publicly to issue statistics on the number and success rates of serious procedures such as open-heart surgery. We soon may see a movement toward requiring individual residents and attendings to make the same sort of disclosures about their individual records.

In sum, it should be apparent that there has been a steady movement toward the requirement that the patient play a decisive role in his own treatment. This movement has placed an obligation upon the physician to provide the patients with sufficient information and opportunity to make choices about his treatment. That obligation strengthens the patient's right to the truth about his condition and also strengthens his right to refuse treatment. At the same time, there is no doubt that a greater burden is being placed upon the physician. The physician's role qua physician to treat illness is taking second place to his fundamental being as a person obligated to respect and recognize the dignity and self-determination of his patients.

Chapter 3

Personhood and the Right to Life

DEFINING DEATH

The recent convergence of technological, medical, and social factors has lent a new interest and urgency to philosophical problems about the ascription of rights, particularly the right to life. Our technology now enables us to keep persons' bodies functioning even after they apparently have irreversibly lost consciousness or have been reduced to states that seemingly no longer include cognitive abilities. Until recently, irreversible cardiac arrest was the sufficient and necessary legal criterion for the determination of death; but now the point at which it could be properly said that a person has died has become unclear.

At the same time that our life-sustaining technology developed, our ability to transplant vital organs and our need for those organs has made the problem of defining "person-death" urgent. The removal of vital organs from a living person presents ethical problems of a different sort than the ethical problems presented in considering their removal from dead persons. After all, in most circumstances, to remove vital organs from a living person is to kill that person.

We see parallel problems about determining personhood when looking at the beginning of human life. For various reasons, the issue of abortion has reemerged as a philosophical, medical, and social problem. Like the adult in deep coma, the conceptus, zygote, and fetus are "less than whole" human beings for at least part of their prenatal existence, and, like the adult in deep coma, the question of what rights they possess, if any, looms large. Just as the issues surrounding the definition of person-death are made urgent because of our need for organs, so the issues surrounding the rights of the fetus are becoming urgent because of our increasing abilities to make use of fetal tissue for transplant and to make use of fetuses for research.[1] Compounding the need to examine the criteria for personhood are questions that have recently arisen about the use of anencephalic infants as organ donors.

Raising the abortion issue often makes people throw up their hands in despair of finding answers. But here too, we are optimistic. There has been a great deal of recent intellectual attention to the issue, and that attention is enabling us to move away from purely emotional approaches and move instead towards a dispassionate clarification and analysis of the underlying theoretical questions. Those who are very optimistic believe that the theoretical work will eventually lead to intellectually indisputable conclusions about the morality of abortion.

We have bundled the issues of person-death and abortion together because they share the same fundamental questions about personhood and the right to life: "What is a person?" and "What are the necessary preconditions for having a right to life?"

The problems are very complex, the more so because coherent attempts to analyze the concept of "rights" and the concept of "person" are quite recent. Because of the complexity of the problems, we must warn the reader that this is the most theoretical and the least "clinical" of the chapters in this book.

The Right to Life—Positive and Negative

The term "right to life" is ambiguous. Drawing on our discussion of the categories of rights in Chapter 1, we may say that the right to life is sometimes looked upon as a *general negative* right, and sometimes looked upon as both a *general negative* and a *general positive* right. A politically conservative interpretation of the right to life would take the position that it is solely a general negative right. That is, a right to life is solely a right not to be killed. In this view, the obligation of government is only to attempt to protect persons from being killed by other persons. It does this by providing a police force and an army to protect persons from being murdered, and by imposing threats of punishment in order to discourage persons from killing other persons. The government has no other obligation in regard to the protection of life, such as providing medical care. Persons have the right to try to earn money to pay for medical care, but have no general right to the care itself.

A politically liberal interpretation of the right to life takes the right both as a general negative right not to be killed and a general positive right to be kept alive if one is in danger of dying. In this interpretation, the government not only has an obligation to protect persons from being killed but also has an obligation to attempt to keep persons from dying. Clearly, in this interpretation, the government has an obligation to provide at least emergency medical care for those who are in danger of dying.

While many believe that there is a *moral general positive* right to life, as of this date there is no federally recognized *legal general positive* right to life in the United States. However, there are instances in which persons may have a *legal special positive* right to life. For example, municipal or state hospitals are usually set up with a mandate to provide at least emergency lifesaving treatment to all persons. Similarly, a private or voluntary hospital may be given an exclusive privilege to provide paid treatment in a particular community. The hospital may also be given public funds and tax exemptions. In return for these special rights the hospital may be required to provide at least emergency lifesaving treatment to all persons, regardless of their ability to pay for treatment.

The question of whether the right to life is a positive right is crucial to the issue of abortion, as we shall see later on. In the meantime, unless we specify otherwise, we will use the phrase "right to life" to refer only to the negative right.

Person

The term "person" is also ambiguous. At the least, it has different connotations in philosophical, legal, and psychological contexts. We will take the term here as synonymous with "possessor of a right to life." That captures the part of the meaning of the term that is crucial to our discussion and will allow us to treat our two questions "What is a person?" and "What are the necessary preconditions for having a right to life?" as one.

Whole Brain Criteria

Until recently, cardiac arrest was almost inevitably followed by the cessation of all other vital functions. Now technological developments enable us to keep the person's somatic functions going long after the brain has suffered severe damage resulting from the deprivation of oxygen or from trauma. As we have said, this ability to keep comatose persons "alive" has come at a time when we also have a general need for transplant organs. The viability of donor organs is greatly enhanced the longer they remain "supported" by the donor's somatic system; thus the comatose would be a good source of organs for transplant.

The costs, financial and other, of keeping comatose patients "alive" are high. They are a drain on hospital resources, on hospital staff and, very often, an emotional drain on family and friends. Our increasing ability to sustain the somatic functions of such persons for longer and longer periods is exacerbating these problems.

Most of those reasons just described are utilitarian considerations that have brought us towards reconsidering the "status" of comatose patients. However, utilitarian reasons alone wouldn't justify the reclassification of certain patients as dead persons. There are theoretical reasons to justify the reclassification, as we shall see.

All of these factors[2] have resulted in ongoing attempts to supplement the traditional (irreversible cessation of cardiac and respiratory function) criteria used for determining the death of persons. In 1968, a committee formed at the Harvard Medical School issued a new set of suggested criteria for defining death.[3] Many states have accepted the criteria, or variations of the criteria, as legally binding. In 1981 the President's Commission for the Study of Ethical Problems in Medicine and Biomedical and Behavioral Research[4] formally proposed that the federal government and the states adopt a version of brain-death criteria for person-death known as the Uniform Death Act.[5] However, at this writing, not all states have accepted "brain death" as a criterion for death, and we have a bizarre situation in which a person who is legally dead in one state is legally alive in another.[6]

The philosophical claims underlying the acceptance of whole brain activity related criteria for death are that (1) the essence of a person is not to be found in the characteristics of his body or in his somatic functions, but, roughly speaking, is found in his consciousness or personality; (2) the presence of consciousness in a person is inextricably tied to his having a functioning central nervous system; (3) irreversible cessation of a person's brain activity is a sufficient condition for the irreversible destruction of his consciousness; and (4) irreversible loss of consciousness is a sufficient condition for the "loss" of at least certain rights, including the right to life.

We may better visualize the reasoning by imagining what the results of a brain transplant would be, if it were possible to do one: If Percy were to have his brain removed and replaced by Mary's brain, we would think that Mary had received a "body transplant," rather than that Percy had received a brain transplant. The essence of Mary's personality, tied to her brain, would now "reside" in Percy's body. For example, upon awakening after surgery, the brain recipient would have the physical characteristics of Percy, but would have the essential personality of the brain donor, Mary. When asked "his" name, the recipient would reply "Mary." Asked to give a "social history," the biography would be that of Mary, probably given in a deeper voice. As for Percy, his fate is tied to whatever was done with his brain. If his functioning brain was destroyed, then Percy was killed, even if his body functions were kept "going."

Using electroencephalograph readings as indications of brain activity,[7] the Harvard criteria and similar criteria require a carefully confirmed determination that there has been an irreversible loss of all central nervous system activity. After confirmation, the person will be declared dead and, under ordinary circumstances,[8] all interventions that sustain somatic functions may be withdrawn. If the "neomort" is to be the source of organs, then somatic life-sustaining interventions may be continued until a recipient is found for the organs, at which point the organs may be removed.

Neocortical Death

We are already seeing claims that these criteria, which demand that there be total irreversible cessation of brain activity before death can be declared, are too strict. It has been suggested that a finding of irreversible destruction of the *upper* brain should be sufficient indication that death has occurred.[9] It has also been suggested that the absence of an upper brain in anencephalic neonates is an indication that the anencephalic is not a person. The thinking here is an extension of the belief that personhood has to do with consciousness, and is an attempt to refine the concept of what aspects of consciousness are essential to personhood. The underlying claim is that the essential characteristics of a person are to be found in the functions of the upper brain, and the mere presence of reflexes, spontaneous respiration, and even simple sensation are not sufficient indications that a person is "alive."[10]

We may visualize the reasoning here with another transplant analogy: Percy suffers from a congenital inability to feel pain.[11] The cause has been localized to an untreatable malfunction of his thalamus. Mary offers to donate her thalamus for transplant. On the assumption that Percy's inability to feel pain is associated with the thalamus, and that the thalamus is not immediately involved in any "higher" consciousness activity, Percy should find this surgery acceptable. He will keep his upper brain and thus keep most of his own original "person," although certainly his personality will begin to change somewhat now that he has the ability to feel pain.

While the upper brain criteria for death seem to have theoretical merit, they are fraught with many difficulties. First, we are faced with very deep philosophical problems when we try to specify precisely which mental states are necessary to a person's identity. The whole brain death criteria avoid those problems[12] by not trying to be specific about the mental states that define a person. Most traditional philosophical theories of personal identity speak about characteristics such as "having a consciousness of self," and/or "having memories of one's personal history," as the essence of a person's identity; but the criteria are still arguable. It seems reasonable to propose that we can be certain that whatever the criteria are that do establish personal identity, they must be associated with the *upper brain*. On the other hand, it could be argued that once we try to "localize" personhood, we have started on a slippery slope that will lead us towards declaring some persons dead who in fact are not. Secondly, and along the same lines, our knowledge of the relationship between brain function and consciousness is primitive at present. Our ignorance is not just a matter of our current lack of knowledge about the "mapping" of neurons, but is a reflection of many conceptual difficulties that have to be solved before we have a satisfactory theory of consciousness. We are certain[13] that consciousness ties to central nervous system activity, and certain that the total irreversible cessation of brain functioning means the end of the person's life (at least in this world). We can be pretty certain that personality is specifically tied to the upper CNS, but, for example, we know very little about the ability or lack of ability of the brain to "regenerate" lost consciousness functions. Nor do we know how to localize most types of consciousness to specific parts of the upper brain if, in fact, the various consciousness functions *can* be localized.

Thirdly, under the whole brain death criteria, once artificial life support systems are removed, somatic functions such as respiration cease very quickly or simply do not exist. Under the suggested neocortical criteria, many somatic functions of the "neocortical dead" can continue unsupported by technological means. An acceptance of neocortical criteria would mean that we will have "cadavers" that respond to stimuli, and that sometimes have spontaneous respiration. Whether society will ever accept that such "persons" are dead is a question that cannot yet be answered, but one that will certainly emerge as an issue if the upper brain criteria ever come under widespread consideration.[14]

There have also been religious objections to any brain death criteria, and these raise a special set of problems. Some Orthodox Jews, for example, continue to take irreversible cardiac arrest as the *only* acceptable criterion for the determination of death.[15] It is difficult to see a resolution to this particular problem of trying to accommodate a religious principle to a secular principle. In New York, for example, proposals were made to allow individual exceptions to a declaration of death based on whole brain criteria, or to permit a secular declaration of death, but to force hospitals to maintain the brain-dead person if the family or patient so desires. But at stake are the costs of maintaining such "patients" in hospitals, and it is questionable whether insurance companies can be forced to pay for the continued "care" of such "patients." In addition, there are legal problems that may hinge on the declaration of death. For example, it could be easily imagined that the perpetrator of an assault could be charged with assault if his victim was an Orthodox Jew, and homocide if the victim was of another religion.

The most fundamental ethical issue raised by the suggested part-brain death criteria has to do with the ascription of rights and brings us back to the discussion of the beginning of this chapter: What is it that gives an entity a right to life? For even if we had a satisfactory theory of the relationship between brain activity and consciousness, and could pinpoint what neural functions are tied to what aspects of consciousness, we would still have to solve the problem of the connection of consciousness to rights. That is, we would have to know what levels and types of mental activity of consciousness are relevant to the possession of rights, particularly the right to life.

ABORTION

Human Being

There may be a temptation to say that simply being a human being is a sufficient condition to be entitled to rights, particularly, the right to life. However, the term "human being" is ambiguous. Sometimes it is taken to be synonomous with "person" and, in that reading, it simply begs the question to say that "All persons have rights because they are human beings." Sometimes the term "human being" is used in a biological sense and "human being" is taken to mean "member of the species Homo sapiens." The claim that all members of our species have a right to life simply by virtue of their being members of our species presents major difficulties. For example, we certainly don't mean that claim to apply to human cadavers, although they are biologically identifiable as members of Homo sapiens. Of course that may seem a cavil, and it may seem obvious that the claim is really about "living members of the species Homo sapiens." But, as it stands, even that modified claim is inconsistent with the current acceptance of brain death as the criterion for person death. After all, as we have said, the brain-dead person is, in some sense, still alive. Such a "neomort" has cellular division, probably has nutritive capacities and, as well, may have reproductive capacities. Respiration and heart function are also present. The fact that some of these functions must be supported by external means seems irrelevant to the issue of rights. After all, a fully conscious human with spinal injury might need the same external support, yet we would not deny him the right to life.

Of course, one could continue to claim that being a living Homo sapiens is a sufficient condition for having a right to life and so deny the validity of any brain death criteria. However, that claim still leaves us with a fundamental question about the "moral relevance"[16] of being a member of our species to being a possessor of rights. That is, why does the biological fact of species membership entitle something to a right to life? And if the claim that being a living Homo sapiens is a necessary as well as a sufficient condition for the possession of a right to life, we are left with more questions. For example, why shouldn't the invertebrate animal which has CNS function equivalent to that of the brain-dead human also have a right to life? In fact, the very claim to membership in our species may not be made on solely biological grounds. The question of whether a fossil hominoid is to be categorized as a "Homo sapiens" rather than as a member of another species may depend more on paleontological evidence about his social life (and hence, by extension, his intelligence) than it does on his bone structure.

A basic principle of fairness requires that we treat two things that are alike in morally relevant factors as equals. For example, given two candidates for a medical residency who have equal qualifications, it would be wrong to discriminate against one because of his skin color. Skin color is an irrelevant moral factor in this context. Similarly, given two animals with the same level of consciousness, it would seem to be unfair to give the one rights simply because he is a Homo sapiens and to deny the other rights simply because he is a member of another species.

It seems that we must examine criteria other than species membership in order to decide the question of personhood. The direction appears to lead towards examining the relationship of consciousness to personhood.

These are also the issues that are directly relevant to the abortion issue. Very roughly speaking, the level of consciousness of the fetus is in some senses similar to that of demented adults and to that of some nonhuman animals. In fact, it doesn't even appear likely the fetus has *any* consciousness in its early stages of development.

Much of the dispute between those opposed to abortion and those in favor of permissible abortion has been confused. Those arguing for continued legalized abortion often miss the basic point of those opposed to abortion.

Commonly, arguments given to support abortion are given in terms of the burden that having a child will place upon the pregnant woman, and in terms of a concomitant right of the woman to refuse that burden. The burden may be financial, emotional, or physical, or all three. An argument is also often offered that the pregnant woman has the right to do whatever she wishes with her body. This argument claims that there is simply no need for the woman to provide any justification if she decides to have an abortion. Those opposed to abortion usually argue that such reasons and arguments cannot justify what they look upon as the killing of an innocent person (the fetus).

If, in fact, the fetus is a person, many of the proabortion arguments mentioned above are faulty. For example, it would be hard to justify the killing of an innocent person just because he is a financial or emotional burden. If my infirm adult relative were a financial and emotional burden on me, I would not be justified in killing him.

That problem of justification persists even if the continued life of the fetus is a threat to the life of the pregnant woman. That is, in those cases in which the fetus must be killed in order to save the life of the pregnant woman. I *may* be justified in killing an innocent person who is a threat to my life, although that is certainly debatable. However, it is unlikely that I could justly demand that a third person kill the innocent person in order to save my life. Nor would that third person be justified in acceding to my request or demand. Thus, self-abortion *may* be justifiable if the fetus is a threat to the life of the pregnant woman, but abortion by a third person—the physician—may not be as easily justifiable.

But, again, it must be stressed that those are issues that arise only *if* the fetus is a *person*. This is the fundamental question to be answered.

History

In general, Western tradition has never given the fetus the status of full personhood. For example, abortion as well as infanticide were widely practiced and accepted in ancient Greece. In the dialogue *Theaitetos,*[17] Plato portrays Socrates as comparing his own philosophical work to his mother's work as midwife. He describes his role as analogous to that of the Greek midwife—examining ideas as if they were newborn babies, and keeping the good ones and discarding the bad ones. The comment was made as a clever analogy that his audience could easily understand, not as a statement meant to shock an audience that already accepted infanticide.

The Hippocratic Oath requires the physician to swear that he will not administer abortion-causing drugs. However, that might very well be interpreted as a statement about the status and methodology of the Hippocratic physician himself rather than as a reflection of a belief about the moral status of the fetus. That is, just as "taking up the knife" was beneath the dignity of the Hippocratic physician and so to be left to apothecaries, so abortion was to be left to midwives.

Many early Christian thinkers appeared to accept viability as the point at which the fetus had an independent moral status.[18] That also appears to be true of

Western secular legal thinking. Viability as the morally significant point of development of the fetus became accepted in common law. The later influence of Aristotelian thinking upon the Roman Catholic Church in the twelfth century led to a change in belief about the status of the fetus. Aristotle had held that the development of self-movement in the fetus was an indication that the fetus had developed a mind. In Catholic thinking, this "quickening" became evidence of ensoulment, and thus the fetus was considered to be an ensouled person when the pregnant woman could feel the fetus move. After that point, abortion would be considered to be killing of a person. At the same time, the Aristotelean-natural law theories about essence and development were taken to imply that it was also wrong to interfere with the developing of the conceptus into a person. The theories were also used to bolster the belief that there was a duty for married persons to produce children. These beliefs remained doctrine until the late nineteenth century when the Church accepted the moment of conception as the point of ensoulment and, hence, the point at which abortion became homocide.

Roe v. Wade

United States law kept the common law tradition of viability as the morally significant point in fetal development. However, even viability did not guarantee full legal personhood to the fetus. Thus, for example, the national census never included fetuses in its count of the population, and the killing of a fetus was never considered to be a homocide.

However, there were a number of state laws against abortion which appeared in the late nineteenth century, and these were the laws that the Supreme Court was asked to consider in Roe v. Wade.

The Supreme Court, in overthrowing these laws forbidding abortion in Roe v. Wade, used the following reasoning:

First, the court reiterated the common law position that the fetus has never had the legal status of a person.

Second, the court held that the nineteenth century laws against abortion were created to protect the pregnant woman, not the fetus. That is, before the development of aseptic techniques, abortion was a dangerous procedure for the woman. Therefore, the states, using their *parens patriae* powers to protect persons from harming themselves, prevented women from undergoing the procedure. In the twentieth century, however, the situation has changed. Abortion, in the early stages of pregnancy at least, is a less risky procedure than childbirth. There are statistically more deaths per thousand in childbirth than there are in early abortion. Thus, the Supreme Court held, it was no longer necessary to protect the woman by an absolute prohibition of abortion.

However, the court made some important provisos. While recognizing a person's right to privacy or self-determination, it limited that right by reiterating the *parens patriae* power of the state to protect persons from harming themselves, and required that abortions be performed by qualified medical practitioners.

The court also said that legal limits on late stage abortions are permissible, and gave two reasons: First, that there is a significant increase in risk to the woman in late stage (after the first trimester) abortions. Second, falling back on the common law acceptance of viability as a morally significant stage in the development of the fetus, the court held that since "the [late stage] fetus presumably has the capacity for meaningful life outside the mother's womb"[19] a state may "go so far as to proscribe abortion during that period, except when it is necessary to preserve the health or life of the mother."[20]

In summary, there are important points to note in the Supreme Court decision in Roe v. Wade.

1. The court did not take away any existing legal rights of the fetus, it simply upheld the traditional view that the fetus had no legal rights.

2. The court did not claim that the late-stage fetus has full personhood and full protection of the law. It permitted killing of the late-stage fetus under certain conditions—for example, if the fetus was a danger to the pregnant woman. If the late-stage fetus were a full person, it could not be killed even to save the life of another person. If it had full protection of law, then certainly a trial or inquest would be required to see if the abortion was, in fact, justified.

Philosophical Issues

While it is important to know something about the history and present legal status of abortion, that knowledge alone doesn't settle the philosophical issues about the personhood and moral rights of the fetus, if it has any.

It will be useful to begin discussion of the philosophical issues by briefly describing what our philosophical theories state or imply about the possessors of rights, particularly the right to life.[21]

Autonomy Theory (Kantian)

Strictly speaking, Kant's original moral theory entails that only entities that are capable of acting for moral reasons (in modern terminology, "moral agents") have any rights and thus are "persons." In Kant's view, to be a person one must have a "rational will." To have a rational will is to have the ability to have a conception of "obligation," to have the ability to try to act according to that concept, and to have the ability to be a "lawmaker." Those abilities require a rather high level of conceptual ability. An entity that lacks a rational will may behave "morally" because of its instincts or conditioning, but such an entity is without a rational will, is not a moral agent, and thus not a possessor of rights. For example, a cow that will not kill its owner is behaving in accord with a moral injunction against killing persons, but is acting out of instinct rather than out of a concept of obligation. As far as we know, on this planet, only adult normal rational humans have rational wills. Strictly speaking, then, Kant's theory implies that nonhuman animals, children, fetuses,[22] the very psychotic, the demented, and the brain-dead adult human do not have rights.

Since being alive is a precondition for the existence and exercise of our rational wills, there is a perfect duty[23] not to kill persons. That is, persons have a *negative right to life.*[24]

Social Contract Theory

The implications of early versions of social contract theory about rights possession are similar to Kantian theory. Only those entities that are moral agents have rights. To have rights requires that the entity be capable of understanding the implications of a contract, and understanding the required trade-offs involved in contracting for rights and obligations. Thus, nonhuman animals, children, fetuses, and the very retarded do not have rights.

As we stated in Chapter 1, Hobbes believed that we have an inborn desire to stay alive. More than that, since being alive is the basic prerequisite to pursuing our desires and interests, Hobbes is often interpreted as claiming that *the negative right to life is the most fundamental* of all rights.

Natural Law

The major natural law theorists,[25] Aristotle and St. Thomas, wrote before the term "right" was used. But if we tie the concept of rights to their concept of obligations, we may get some idea of what early natural law theorists would say about rights. Since the adult rational human being is considered the locus of our obligations, we can say that adult rational human beings are the possessor of rights. These rights would function to protect the fulfillment of our essence.

In St. Thomas' view, at least, our essence includes our having an innate desire to stay alive. Moreover our ability to fulfill our earthly purposes necessitates our being alive. Thus we have a perfect duty not to kill ourselves or others. Therefore, it could be said that adult rational humans have at least a negative right to life. In addition, since we have a duty to procreate, and since the essence and purpose of the fetus is to develop into an adult rational human being, it could be said that there are obligations towards fetuses. The Catholic Church takes those obligations to be very strong, claiming that intentional direct intervention to interfere with procreation (the use of artificial birth control) is wrong, and further claiming that the intentional direct killing of a fetus is homocide. We will discuss this issue of the potential of the fetus in depth.

Utilitarian Theory

As we said in Chapter 1, it is not clear that the concept of rights is compatible with utilitarian theory. There is a utilitarian obligation to take the interests of all sentient creatures into account, and, in that sense, all things capable of sensations of pleasure and pain have a right that their interests be considered. Presumably, the rightness or wrongness of abortion would depend on the total consequences that policies of permitted or prohibited abortion would have for society.

The moral agent criteria for rights possession given by Kant and Hobbes seem too strict. For example, they call into question the existence of any rights for those who are not moral agents, including children, the very psychotic, and even the moderately retarded. However, moral agent criteria are very much embedded in our legal system of rights. Some theorists have tried to argue that moral agent theories can account for rights of these nonmoral agents. However, they end up with positions that make those rights really "second class" and dependent upon the wishes and desires of moral agents who have "first class" rights. For example, such theorists may argue that while the rights of moral agents are fundamental, those of nonmoral agents are "given" to them by moral agents because it would affront the sensibilities of moral agents if these others didn't have rights. Yet most would find it difficult to accept that (for example) a retarded person's right not to be put into needless pain is any less fundamental than the equivalent right of the normal person.

Some recent work in the theory of rights has rejected moral agent criteria for rights, and has tried to approach the issue of rights in different ways. For example, it has been suggested that while there could be no *system* of legal rights if there were no moral agents, moral agency is not a necessary precondition for *having* rights. That is, only moral agents have the ability to set up legal systems that recognize rights and contain mechanisms to deter, judge, and punish violations of rights; but that doesn't entail that one must be a moral agent in order to have a right.

In this point of view, the emphasis is shifted from a consideration of how rights are "set up" in legal and moral systems to a consideration of the reasons that we think it is desirable or obligatory to have rights at all. For instance, it might be said that having a right serves to protect some interest or good of its possessor, so a right must have at least the potential to be useful to those who possess the right. If it is a negative right, it protects the possessor from the encroachment of other persons who have competing interests. If it is a positive right, it can give some positive benefit to the possessor.

Along these lines, it has been suggested that in order to have a right to X, only two conditions are necessary: (1) Moral agents must value that X, or believe that they will suffer at the denial of that X, enough to protect their interest in X with a right; and (2) the principle of fairness demands that any entity, moral agent or not, that can value that X or can suffer at the denial of X, has a right to X.[26]

If the entity does not have the capacity to value X and suffer at the denial of X, then there is no reason, need, or duty to give it a right to X. So, for example, we would have no obligation to recognize that stones have a right not to be put into needless pain. They are incapable of benefiting from such a right and incapable of suffering at the denial of such a right. Too, for example, we would have no obligation to recognize a right to a grade school education for cats because they are incapable of valuing a grade school education, and incapable of suffering from a denial of the education.

We may see the principle behind this argument if we look at how it explains a possible genesis of the negative[27] right to life. First, consider the question: "Why do we think that a right to life is important?" Normal adult human beings are capable of placing a value on living. Moreover, they are capable of placing a disvalue on death. The possibility of being killed, or of being in constant danger of being killed, normally causes us great suffering and anxiety. In order to protect our lives, at least against those who can be swayed by threats, we recognize a legal right to life with accompanying sanctions and threats of punishment against those who would violate the right. So far, that account of the right to life is similar to the accounts of the function of a right to life given by Hobbes' social contract theory, Kantian theory, and natural law theory.

However, this account does not limit the possession of the right to moral agents.

Under this interpretation, the two necessary conditions for possessing a right to life are (1) the capacity to have a conception of what it means to be alive; (2) the capacity to place a value on being alive and place a disvalue on dying.

Having a capacity to have the requisite concepts of "life" and "death" does not necessitate having any profound metaphysical theories about "the meaning of life." It is sufficient to have some concept of what it means to have ongoing experiences and what it would mean for one's experiences to come to an end. Similarly, having the capacity to desire to live is to be capable of having a desire to continue to have experiences and to be capable of disvaluing the possibility of having all of one's experiences end.[28] In order to have the capacity to have these concepts an entity needs some consciousness of "self," some ability to think *about* his experiences,[29] and some ability to think about his future.

On the other hand, having the necessary concepts of being alive and a capacity to desire to stay alive means more than simply having a set of instinctual responses which tend to be useful in keeping one alive. So, for example, an earthworm may have sets of innate responses that tend to keep it alive, but almost certainly the earthworm doesn't have concepts of life or death or self.[30] The earthworm may suffer pain at stimuli that tend to be dangerous to its continued existence, and it may "care" enough about those *specific* stimuli to try to avoid them, but it doesn't have the self-reflective or cognitive capacity to care if it lives or not.[31] That is, its desire is to be free of the pain that it is feeling, rather than a desire to stay alive.

This analysis of the right to life is at least somewhat in accord with our earlier discussion of the definition of person death. The irreversible loss of consciousness in the "brain-dead" person permits us to say that he no longer has a right to life. The brain dead do not suffer from the denial of that right, even though the denial of the right will likely mean the termination of their somatic life. We do not think that they are unjustly denied something if their somatic life is ended. Too, our discussion of upper brain criteria for death in this chapter, and our discussion of patients in chronic vegetative states in Chapter 4, "Euthanasia," indicate that the presence of simple consciousness alone does not seem to imply the presence of a right to life. The objections to the suggestion of neocortical criteria for death center around our difficulties in determining how brain activity relates to "higher" consciousness. Thus, we have difficulties determining the levels of consciousness in adult humans who have suffered some insult to an already developed CNS.[32] Theoretically, though, if we could establish that a patient had irreversibly lost the capacity to care about his continued existence, then it might be justifiable to say that he no longer has a right to life. That might be true even if he still retained a capacity for simple sensations of pain and pleasure. It should be pointed out that this line of thinking is also in accord with a claim that some nonhuman animals may have some rights, such as a right not to be put into needless pain, even if the animal does not have a right to life. That is, it is in accord with the belief that it is permissible to painlessly kill nonhuman animals. Most people do not believe that we have a moral obligation to recognize a right to life in the full sense of the term[33] to "lower" animals with simple nervous systems such as earthworms, although such animals likely have enough consciousness to have pain sensations and thus may have some rights in regard to the infliction of pain.

The Human Fetus

It is highly probable that the human fetus hasn't the capacity to have those concepts of life or death of which we have spoken. If we believe that there is a relationship between the possession of a central nervous system and the possession of consciousness, then we can say that the fetus doesn't even have the capacity for simple sensations (such as pain or pleasure) in its early stages of development. In those early stages, the nervous system is not complete enough for sensation to be present. Even in later stages when the central nervous system has developed more completely, it is unlikely that the fetus could have a concept of self and a concept of what it would mean for the self to continue to exist, or to cease to exist. The capacity to possess those concepts requires more than a well-developed "empty" brain, it requires having had experiences more varied and complex than is possible for the in utero fetus. The human fetus and nonhuman animals with their lack of capacity and/or lack of experiences must be contrasted to the unconscious or sleeping adult, the very psychotic adult, the psychotic suicidal adult, and the mildly retarded adult. These all have at least the capacity to have the necessary concepts and desires. And while the psychotic suicidal adult may not seem to have a desire to continue living, there are other reasons why it would be wrong to kill him.[34]

Under this admittedly complex and controversial analysis, the fetus does not have a right to life. On the other hand, the fetus may, in its later stages of development, have a right not to have pain inflicted upon it. More controversial, the analysis implies that normal human infants and profoundly retarded humans also do not have a right to life since they do not have the requisite capacities or desires. We will come back to speak of these problems later in this chapter.

Potentiality

However, some believe that any analysis of the abortion issue must consider the moral significance of the fetal *potential* for personhood. They might say that, unlike the very demented or comatose adult, the fetus has the potential to become a person. While the adult in irreversible coma and the fetus both have less than whole consciousness, the comatose adult has lost his capacity and potential and is "on the way out" while the fetus is "on the way in." It is argued that the fetus should be allowed to realize that potential.

The concept of fetal "potential" raises a new set of problems.

First, not *all* fetuses have the potential to reach personhood. Some fetuses, such as the anencephalics, certainly do not. Other fetuses that have major defects of their central nervous systems that will greatly limit their cognitive abilities also probably don't have that potential. Thus, any argument against abortion that rests on the fetal potential for personhood cannot be used to limit *all* abortions.

Second, even the normal fetus will not develop into a full-fledged person without help. Human beings are not like amoebas, which are self-sufficient from the moment of their "conception." The presence of potential in a fetus doesn't guarantee that the potential will be actualized if the fetus is simply left alone to develop. It is estimated that over 20 percent of human fetuses will spontaneously miscarry if "left alone" without medical intervention. The human fetus needs a protective and nurturing environment before it can become "viable." Even after birth, the newborn needs assistance before it can live unaided, much less have its "potential" for full personhood realized. Thus, those who argue that the fetus must be "allowed" to develop into a person are really committed to the view that the fetus not only has a *negative* right to life—a right not to be killed—but also has *positive* rights, which include a right to be kept alive and a right to have its potential for personhood actualized.

Third, it is not clear at what developmental point it makes sense to say that a potential person has come into existence. Thus it is not clear at what point it makes sense to say that there is now a duty to actualize the potential person. Some have claimed that a potential person comes into existence at the point of implantation, when the fertilized egg has its individual set of chromosomes and can no longer split into twins. But that point seems arbitrary. With the development of cloning techniques it may be possible to claim that any living cell in the body is a potential individuated person. If there is a duty to help actualize potential persons, does that mean that cloning should be encouraged?

The implication of natural law theory is that the presence of a fertile male and fertile female creates the potential for procreation of persons. Does that mean (as, in fact, Catholic moral theology claims) that the presence of a mature egg in the married woman imposes at least some duty upon the couple to fertilize it in order to help actualize a potential person? Should couples take fertility pills that have the side effect of producing multiple births in order to actualize more persons? Should all birth control devices be banned since they interfere with the actualization of potential persons? These questions may seem absurd; nevertheless they throw doubt on the usefulness of potentiality as a criterion for determining our moral obligations in regard to abortion.

Still, at first glance, there may seem to be something compelling about the belief that it is wrong to interfere with the development of the normal fetus, and that it is wrong not to help the fetus actualize its potential. Part of this belief may be based on an assumption that a fetus is being both harmed and deprived of some good if it is aborted. Sometimes those opposed to abortion will phrase this belief in the form of a question: "Aren't you glad you were born and so isn't it wrong to deny someone else (the fetus) his potential chance at life?" But, of course, whether or not I am now glad that I was born, the question itself would make no sense if I had *not* been born. If I had not been born, I would not have been around to regret the fact. It only makes sense to say that someone was harmed or deprived of some good if he actually is or was around to be harmed or deprived of that good. A nonexistent person is not really missing anything. It is true that we may justifiably feel sad that an existing person's potential was not actualized, and that person may justifiably believe that harm was done to him if his potential was intentionally interfered with. But, for example, it would make no sense for me to feel sad that my nonexistent sister never realized "'her' potential" or feel that she is harmed because she never existed, or feel that she is being deprived of the pleasure of discussing medical ethics with me.

Of course, the fetus, unlike my sister, does exist and thus already is something that (if it is normal) does have potential to be a person. Still, unlike a medical student caught in circumstances that don't allow him to realize his potential to become a physician, the fetus doesn't actually have the capacity to suffer from his deprivation.

In ordinary circumstances, having the potential to be an X does not give someone the same rights as an actual X. For example, the medical student who is a potential physician does not have the same rights as an actual physician.

Moreover, while it is true that it would probably constitute harm to intentionally interfere with the medical student's potential to be a physician, it is not clear if any specific person has an obligation to help the student actualize his potential. As we have said, the potentiality argument implies that the fetus has a positive right to life and a positive right to have its potentiality actualized. Even if that were true, there are questions about who has the obligation to honor those rights.

One question that might be asked is whether the pregnant woman has the sole obligation, or any obligation, to make sure those positive rights are supported. In ordinary instances, the legal burden of fulfilling general positive rights[35] is shared by society at large and not placed entirely upon any individual member of society. Thus, if a starving and homeless person has a right to food and shelter, the cost of that food and shelter is shared by society from tax revenues. If he has a right to have his potential actualized by being given an education, or job training, that burden is shared by society. No one individual has the obligation to provide all of the food and shelter necessary for the homeless and hungry person. Analogously, it might be asked why the pregnant woman should be forced to provide all of the "food and shelter" for the fetus. Reasons might be given to support her obligation. For example, it might be claimed that she is the only person who can provide the life needs of the fetus until the fetus is "viable," i.e., able to live outside the womb. However, that is not entirely true. "Viability" is another ambiguous concept. If it is to mean that the fetus can live outside of the natural mother's womb, then by that criterion, viability can occur quite early in a pregnancy now that we have the ability to remove fertilized eggs and transplant or reimplant them. Even in the later stages of pregnancy, the point of "viability" for a fetus is dependent on what technologies are available. Some hospitals may have the technical facilities to save very low-weight and early fetuses, while other hospitals may have a difficult time saving even slightly premature infants.

Whether or not the fetus is viable enough to live outside of the pregnant woman, and whether or not it is an "actual" person or only a "potential" person, it still has to be established that the woman is not within her rights if she chooses to have the living fetus expelled.[36] She may argue that by doing so she is not directly killing the fetus, she is just refusing to take on the sole obligation of keeping it alive. Under ordinary circumstances, no person has a legal obligation to keep another person alive. For example, if I am the only available blood match for a person who needs blood in order to live, I am not compelled by law to donate that blood. There are important special exceptions to that rule. For example, the physician who has taken on a person as a patient ordinarily does incur the obligation to try to keep him alive. These, however, are obligations that have special contractual justifications and are obligations that do not apply to laypersons. There are "Good Samaritan" laws in existence in some places that require laypersons to at least attempt to help victims of accidents or crime. But the requirements of these laws are minimal—they do not demand that the Samaritan put himself at risk or exert "extraordinary" effort in order to help the victim.[37]

It might be argued that the woman takes on moral obligations towards the fetus when she becomes pregnant. That is, in becoming pregnant she has incurred special contractual obligations similar to those incurred by a physician when he takes on a patient. In our terminology of rights, this claim would be that a "special contract" has been created between the woman and the potential person with accompanying special rights and obligations. Certainly this argument is unsupportable in instances where the woman has become pregnant involuntarily, as in the case of rape. There are other instances in which the woman could not be said to have become pregnant voluntarily. For example, a minor female or a very retarded female or a very psychotic incompetent female could not be thought of as having voluntarily become pregnant. Thus, this argument alone could not support the claim that a pregnant woman *always* has an obligation not to expel the living fetus.

Even if the woman voluntarily becomes pregnant, the argument that she thus takes on an implicit and unbreakable "contract" is questionable. First, the claim that a woman can make a contract with the fetus which must not be broken by aborting the fetus depends on that arguable supposition that the fetus as a potential person has the same rights as an actual person. That is, the argument already presupposes that the fetus has positive and negative rights to life.

Infanticide

We have said that the theory of personhood offered above seems to imply that infanticide as well as abortion is morally permissible.[38] It seems unlikely that infants have the capacity to have the concepts required for a right to life much before the age of two years. It should be stressed, though, that the theory implies that they would have rights in regard to pain and other forms of suffering. In fact, any argument offered for permissible abortion has to contend with the question: "What is the difference between killing the living fetus in utero and killing the newborn?" It has been argued that once a viable fetus has been born, it becomes part of our "moral community" and thus has rights. However, such an argument doesn't explain why the newborn, or any entity that doesn't meet requirements for full personhood should or should not be accepted into "the moral community."

We have seen that regardless of the moral status of the fetus in regard to rights, expulsion of a living fetus might be justified as an instance of a woman's right to refuse to take on the total duty of providing for the fetus. If such a fetus is viable, then her action could be seen as comparable to that of a woman who gives up a full-term newborn to the state for adoption.

However, expulsion of the living fetus, even if it results in the death of the fetus, is not the same as actually killing the fetus.

Moreover, while many persons might agree that it is morally permissible to allow severely defective newborns to die, it is unclear how many would be ready to admit that it is permissible to actually kill them (see Chapter 4). There seems to be some growing acceptance that it may be permissible to remove the vital organs of anencephalics (thus killing them) for transplant purposes. There is also some growing public discussion about the personhood of anencephalics. However, the anencephalic infant with its total lack of upper brain and its inevitable poor prognosis for extended life is clearly different from, for example, the profoundly retarded infant with a capacity for extended life.

It is probable that very few persons would accept a claim that it is morally permissible to either kill normal infants or allow them to die. It is likely that any such suggestion would be met with repugnance by most persons. That is not to say that the discomfort or repugnance is justified, and it may be the case that society's emotions simply have not yet caught up with and adjusted to ethical analysis. After all, it is likely that most laypersons still would have a hard time adjusting to the concept of brain death if they were confronted with the sight of a brain-dead person still connected to machines that kept him warm and breathing. We tread on very difficult ground when we try to base morality on emotions alone. In some instances, an emotional response to a question of ethics may be inappropriate and caused by a lack of understanding of the issues. The person who, upon seeing his brain-dead relative supported by a respirator, vehemently refuses to believe that the relative is dead could be an example of that lack of understanding. Such persons may change their emotional responses after being educated a little. On the other hand, an emotional objection to an ethical claim may be an initial sign that there is "something *wrong*" with the reasoning underlying the claim. In either case, the reasons for the emotional response should be examined to see if the response has an intellectual justification.

All in all, the moral status of infants and of the profoundly retarded pose great problems given the stringent requirements for personhood in classical as well as recent theories of rights.

In general, it seems clear that the burden of proof lies with those who believe that the fetus is a person. On the assumption that the fetus is not a person, many of the pro-abortion arguments mentioned earlier in this chapter do have validity. However, the reaffirmation in Roe v. Wade that the fetus is not a person has led to the emergence of a number of other problems. We can only sketchily describe some of these issues in our limited space.

Duties towards the Fetus

There might, for example, be reason to support a claim that *if* the pregnant woman intends to bring the fetus to personhood, rather than choosing to abort, she does have certain forward-looking obligations to that future person. These might include duties not to cause harm to the fetus that will adversely affect its development and, as well, duties to provide proper prenatal care and an environment that will be conducive to the development of the fetus into a full-fledged normal person. As we have said before, it makes no sense to say that a person was harmed by not being brought into existence. However, it does seem to make sense to say that an existing person who is now suffering as a result of avoidable intentional or reckless actions (or inactions) by some person, be that person his mother or an obstetrician, during his gestation could claim that that person was blameworthy, and could claim that his rights were violated.

The law has already confirmed the culpability of physicians for that type of harm, and there are a number of court cases pending which are considering the possibility of holding pregnant women culpable for that type of harm.

The Use of Fetuses and Fetal Tissue

The use of live fetuses and fetal tissue could afford enormous opportunities for medical treatment and research. At this writing, for example, there is evidence that the implantation of fetal brain tissue into patients with Parkinson's disease can arrest the progress of the disease and probably reverse at least some of its symptoms. There is also evidence that the implantation of fetal bone marrow may be useful in the treatment of sickle-cell anemia. In general, the fetus with its undeveloped immune system may be a useful source of organs and tissue for transplantation. Moreover, the study of cell differentiation in the fetus could yield important information about the development of cancers, and, perhaps, eventually yield information that would enable us to learn how to regenerate organs, rather than having to depend on transplants.

Given then that the fetus is not a person, are there any moral limits to the use of fetal organs and tissues for transplant, and the use of living fetuses for research? Could a woman sell her fetus for such use? Could she become pregnant with the intention of either donating or selling the fetus for such uses?

Abortion, Childbirth, and Health

Since public monies for health care are allocated for necessary medical procedures, should these monies be used to pay for abortions in cases in which bearing a child would not be a direct health risk to the pregnant woman?

Conversely, should public monies be used to pay for artificial insemination for infertile couples? Should the inability to have children be considered a major health problem? If these procedures are to be provided at the public expense, does the state have the right to demand psychological and genetic screening of the prospective parents?

Forced Abortion

Could the state be justified in forcing abortions in instances in which it is believed that the child would be better off not being born? For example, suppose that a fetus has been diagnosed with AIDS and the pregnant woman refuses to have an abortion. Given that the child, if born, would suffer a short and painful life, would that justify a forcible abortion?

Chapter 4

Euthanasia

We will take as our definition of "euthanasia" the providing of death for the patient for the sake of, or upon the request of, the patient. In doing so, we will include in our discussion the withdrawal of life-support systems and measures with the intent of letting the patient die, as well as the withholding of life-support systems or measures, and the administering of drugs or procedures with the intention of causing death.

It has been said that modern medicine now has the ability to prolong death as well as the ability to prolong life. While "Do no harm" used to mean "Use whatever measures available to keep the patient alive as long as possible," there is a growing consensus that there are cases in which aggressive treatment will cause the patient more harm than good. In other words, there are many who now believe that there are cases in which the patient is better off dead. Moreover, the growing recognition of a right to refuse treatment is including the recognition of a specific right to refuse life-sustaining measures.

While euthanasia has always been an ethical issue in medicine, attempts to bring it to open discussion in order to set legal criteria for its "administration" are quite recent.

ACTIVE AND PASSIVE; VOLUNTARY, NONVOLUNTARY, INVOLUNTARY

We may roughly schematize different types of euthanasia as follows:[1]

	Active	*Passive*
Voluntary		
Nonvoluntary		
Involuntary		

Active: In active euthanasia, a procedure is used that is directly intended to cause the death of the patient. For example, the injection of a lethal mixture of curare and barbiturates.

Passive: In passive euthanasia, a procedure is withdrawn from the patient, or withheld from the patient, with the intention of letting the patient die. Intentionally withholding attempts at cardiopulmonary resuscitation from a patient who has undergone cardiac arrest, removing a respirator from a patient who is not dead, withholding antibiotics from a patient with pneumonia, and withholding lifesaving surgery from a defective neonate are all examples of passive euthanasia.

Voluntary: In voluntary euthanasia, we have the fully informed consent of a competent[2] patient.

Nonvoluntary: Here the patient is not competent. The patient may be incompetent because he is unconscious—such as a patient in coma. He may be incompetent because of his permanent mental state—very senile or severely retarded. He may be only potentially competent—such as a normal infant, severely psychotic adult, or a nonpsychotic adult in mental distress as a result of severe pain or anxiety, or whose judgment is impaired by medication. Of course, there may be mixed causes for the incompetency.

Involuntary: Euthanasia contrary to the wishes of the patient, or without the consent of a competent patient.

INVOLUNTARY ACTIVE EUTHANASIA

It appears obvious that involuntary active euthanasia is impermissible. Involuntary active euthanasia is theoretically compatible with utilitarian theory, and utilitarian arguments might be presented that would appear to justify the practice.[3] However, it is likely that a permitted policy of involuntary active euthanasia would set a dangerous precedent and have unhappy long-term consequences for society. It is hard to imagine a society in which persons could be content with the knowledge that others could go against their wishes, decide that they would be better off dead, and then proceed to kill them.

A more fundamental argument against involuntary active euthanasia comes from autonomy theory. A synopsis of the argument is as follows: "No matter how secure we are in our belief that a patient would be better off dead, we cannot impose that belief upon a patient who prefers to hold on to life. Even if he chooses to live under conditions in which we ourselves would choose not to endure, that must remain his right as an autonomous being.

The social contract argument is similar: "The right to life is achieved by means of the individual's contract with society. With the possible exception of persons who have violated others' rights, only the possessor of the right to life may waive his right."

INVOLUNTARY PASSIVE EUTHANASIA

The primary argument against involuntary passive euthanasia is analogous to the autonomy argument against involuntary active euthanasia. An individual has the right to decide whether or not he wishes to continue living. To withdraw or withhold life-sustaining treatment because we judge that the patient is better off dead is to paternalistically violate his autonomy.

Moreover, in cases in which the patient is already receiving life-sustaining treatment, withdrawing treatment against the patient's wishes, or without the patient's consent, might also be a violation of a contractual right. If treatment was already initiated on the patient, removing treatment against the patient's wishes would be a violation of the contract between hospital and patient or physician and patient. Under ordinary circumstances, a person can presume that once he has become a patient and treatment has been initiated, treatment will be continued. Even if the removal or withholding of life-sustaining means were to be looked upon as the "best treatment" alternative[4] rather than simply as a termination of treatment, we would have a situation in which there was a failure to get informed consent from the patient for this "best treatment" alternative, and thus have a violation of his rights.

"Slow Codes" as Involuntary Passive Euthanasia

At present, many hospitals still have only informal "policies" for "no code" or "do not resuscitate" orders. Sometimes the policy is ad hoc and simply set by the attending physician or house officer on call in the care unit. In some instances, the policy informally permits the physician to simply assume that a patient is incompetent and allows a decision to be made by the family or physician not to resuscitate the patient. In other instances, the practice is rationalized by assuming that even if the patient is competent, discussion with him about a DNR (or other withholding of life-sustaining measures) will cause him great psychological harm.

If no attempt is made to obtain consent from the patient, or to test the patient's competency before letting others make the decision, then such a policy is at best questionable. As legal guidelines get set for passive euthanasia, it is certain that they will require a formal determination of incompetency or mental incapacity before a decision to withhold or withdraw treatment may be made by persons other than the patient. Even those policies that allow a "therapeutic exception"—permission to write a DNR for a competent patient without his consent if it is judged that discussion will cause harm to the patient—will likely require confirmation that discussion with the patient would, in fact, cause serious harm. These practices are mentioned in this section because while they are often presented as nonvoluntary passive euthanasia, they are, in fact, *involuntary* passive euthanasia. The law, reflecting some of its autonomy theory underpinnings, presumes persons to be competent to make decisions until proven otherwise. If no attempt is made to question the patient about his wishes, it can only be assumed that the patient's consent and wishes were thought to be irrelevant to the decision. As for "therapeutic exceptions," it might be assumed that a competent patient's inability to discuss the issue would be an indication of his ambivalence about the issue. That ambivalence might be a reason to refrain from writing such an order.

The practice of calling "slow codes"—giving orders to resuscitate, but with the order for the treatment team to take its time in initiating cardiopulmonary resuscitation techniques—is clearly just an attempt to avoid possible legal difficulties. On those grounds, it is not worth discussing here. We hope that the initiation of clear legal guidelines for "no codes" will soon end the practice of "slow codes."

VOLUNTARY PASSIVE EUTHANASIA

As of this time, some forms of voluntary passive euthanasia are permissible under most state laws and under many professional codes of ethics.[5] However, many of the laws are unclear about the limits and implementation.

There are strong arguments for voluntary passive euthanasia. First, there is the autonomy argument that entails that patients have the right to refuse any treatment. In cases in which the patient is not yet using life-sustaining measures, the principle of autonomy that allows any competent patient to refuse treatment seems broad enough to permit a competent patient to refuse to accept life-sustaining meausres, or resuscitative measures.

While the state has always reserved a *parens patria* power to prevent persons from harming themselves,[6] most courts are tending towards supporting patients' decisions to refuse treatment, or to have life-sustaining measures withdrawn. These court decisions are in accord with autonomy theory[7] and often, as well, in accord with natural law theory. On the other hand, many of the decisions are limited to cases in which the patient is terminally ill. In setting these limitations, there is a reliance on a second argument.

The second major argument for voluntary passive euthanasia was mentioned at the beginning of this chapter; there is a growing consensus that sometimes technology may harm the patient by keeping him alive. That is, there is a belief that while the use of any medical procedures that keep patients alive may have seemed to be called for in the past, we now have situations in which such procedures not only are useless, but may be said to be contraindicated since they will cause suffering to the patient, while not improving his prognosis.

These two arguments ("the right to refuse treatment" argument, and the "best treatment" argument) are distinct, and have different ramifications for the issue of active euthanasia.

EXTRAORDINARY AND ORDINARY MEANS[8]

Some life-sustaining or resuscitative measures are sometimes spoken of as means that are "*extra*ordinary." The terminology derives from natural law theory. While natural law theory and laws derived from natural law theory do not permit persons to harm themselves and do require that persons accept some medical treatment, they do not require that persons use extraordinary means to keep themselves alive. Nor do they compel others to provide those means. Thus, on the one hand, the patient's request for the removal of extraordinary means or his request that they not be used is not construed as an instance of a person intending to harm himself. Similarly, the physician's decision (with the patient's consent) not to use extraordinary measures is not construed as an instance of the physician harming or intending to harm the patient, and not construed as a failure of his duty to treat.[9]

Of course there are problems with defining "extraordinary means." The phrase is usually interpreted as referring to measures that can be used or obtained only with great difficulty, and/or measures that would cause a great deal of suffering to the patient without causing a marked improvement in his prognosis. However, what might be extraordinary in a small hospital in a rural community may be quite ordinary in a large urban hospital. Moreover, what counts as extraordinary may be dependent upon technology. At this date, the implant of a mechanical heart would count as extraordinary. That will likely change in the future. While most physicians would count nasogastric feeding as an ordinary procedure, and many natural law theorists consider feeding a part of basic care and not a medical measure at all, a recent court case using the "extraordinary–ordinary" distinction resulted in the decision that the procedure could be called extraordinary under certain conditions.[10]

Sometimes an attempt is made to define "extraordinary means" in terms of the "burden" the treatment would place upon the patient and upon society. Here again, the criteria to determine "burden" are not at all clear.

There seems to be a general trend in secular legal and moral thinking towards a belief that the distinction between "extraordinary" and "ordinary" means is not useful and thus ought to be abandoned in formulating policy.[11] The same objections hold for those who try to use the concept of "heroic measures." However, the distinction is still used in some religious thinking. For that reason, physicians should be aware that the distinction may be important in the care of those patients who are guided by such beliefs.

WITHDRAWAL VERSUS WITHHOLDING OF TREATMENT

Some physicians take the position that it is permissible to withhold treatment from a patient, but that it is ethically and legally impermissible to withdraw treatment that has already been initiated. Here too, we are faced with a distinction that may not hold up under closer analysis. For example, it is difficult to see a moral distinction between not initiating nasogastric feeding, and deciding not to refill a fluid reservoir that supplies a nasogastric tube already in place. The distinction is not one that has been upheld by court decisions. Moreover, as we shall see, there may be reasons to believe such thinking goes counter to good patient care.

"Do Not Resuscitate"

We have said that passive euthanasia is becoming more and more acceptable in our society. Initial acceptance has usually come by way of legal or institutional requirements that policies be established for writing "do not resuscitate" orders for terminally ill patients whose deaths are considered "imminent." The accepted definitions of "terminally ill" and "imminent" are clearly somewhat arbitrary. Thus, "imminent" and "terminally ill" may sometimes be taken to mean that the patient would not be expected to live more than 6 months even if resuscitated, and sometimes taken to mean that the patient would not be expected to live a year. While the arbitrariness of the definition may not present much of a problem in cases in which the patient is in the end stages of a disease, it does present a problem when the patient may be expected to live a few years after the diagnosis has been made.

As is typical of present-day health care policies, the trend in "DNR" policies is towards protection of patient autonomy and self-determination. The following suggestions for dealing with a DNR order for a competent patient are much in accord with this general trend. Most of the points mentioned are also applicable to the withholding and withdrawing of other forms of life-sustaining treatment.

The possibility often arises that termination of treatment will probably become an option in the care of certain patients. In the clearest of these cases, the patient is in the hospital with an incurable terminal illness and his death is imminent. In such cases the patient should be given his diagnosis and be informed that termination of treatment is an option open to him. It is hoped that the physician will not leave it at that. The physician should offer support to the patient, give the patient time to think about a decision, and make himself available for further discussion with the patient. A liaison team should be readily available to speak with the patient. These may include psychiatrists, psychologists, social workers, and pastors. Of course, the patient should be allowed to discuss his situation with his family and friends, and his family ought to have the opportunity to speak to the liaison providers. However, the decision must rest with the patient himself as long as he is competent. In fact, the patient may request that his family not know his prognosis. That request should be acceded to. However, if the physician believes that the patient may soon become incompetent, he may want to discuss the issue of telling the patient's family with the patient.

The patient should sign a witnessed statement. The patient's request and a record of discussions with the patient should be documented in the case record. The form should be attached to the case record. Of course, the whole treatment team should be notified of the decision. Many difficulties that often arise could be avoided if the attending physician made a practice of discussing these decisions with the entire health care team involved in the care of the patient (particularly since they are the ones who will most likely have to implement the decision). If the patient remains competent, there should be a time restraint on the order. The order should be subject to renewal after a short period. While some might argue that constantly asking the patient about such a decision will provoke anxiety, it would seem that the patient should still be reminded that he has the right to change his mind. The question of renewal could be brought up when the physician has his regular case discussions with the patient.[12]

If the patient has given written consent and becomes incompetent then, unless his prognosis changes for the better, his wishes should be acceded to. A counter-request by the family afterward should not automatically negate the patient's wishes.

A patient might request that he be resuscitated. In such an instance, his wishes should be noted in the case record and, perhaps, he should be allowed to sign a form to that effect. For patients who later become incompetent, much difficulty could be avoided if their requests for or against treatment were documented.

The patient who is indecisive should not be pressured into making a decision and, presumably, patients who have been asked and have not formally made a decision should be resuscitated.

Other Measures

The issue of withholding cardiopulmonary resuscitation ("CPR") for competent patients has been comparatively easy to deal with. The techniques of implementing CPR are discrete and thus easy to define as separate and limited interventions. The procedures of CPR and the consequences of withholding them are easily explained to patients. Cardiac arrest itself is a discrete event and the withholding of CPR will immediately result in the death of the patient. Also, the cardiac arrest usually comes about as a direct consequence of the patient's primary illness rather than as a complication indirectly related to the primary illness. Issues about the withholding and withdrawing of other measures are not as easily dealt with. (It should be pointed out that most DNR policies are quite specific and refer only to the withholding of CPR. Often they explicitly state that they are applicable only to CPR. Thus, even if such a patient has a valid DNR order in his chart, that doesn't entail that other measures can be withheld or withdrawn.)

For example, it is not uncommon for terminally ill patients to develop secondary infections, such as pneumonias. These infections are often easily treatable with the help of antibiotics. Such a patient may be diagnosed with a terminal illness and yet not be thought of as in "imminent danger of dying directly as a result of the terminal illness." Thus, for instance, a patient may be terminally ill with Alzheimer's disease, yet able to continue living for years with minimal care. An untreated pneumonia developed by such a patient will probably cause the patient to die in a short time. Should such a patient be treated for the pneumonia if he does not want treatment?

The general trend that recognizes a competent patient's right to refuse treatment seems to be encompassing the right to refuse interventions such as antibiotics or respirators. But the matter is by no means settled. Again, while the withholding of CPR will result in immediate death, the withholding or withdrawal of many other modalities may involve a period of pain and suffering for the patient before he dies. Moreover, the persistence of conscious or semiconscious attempts to distinguish between "ordinary" and "extraordinary" measures may influence attitudes toward which measures health care providers are willing to withhold or withdraw. Thus, they may be more amenable to withholding or withdrawing respirators than withholding antibiotics, believing that respirators are extraordinary and antibiotics routine.

Certainly, if a patient requests that treatment be withheld, the physician should discuss the implications of that decision with the patient. A patient may have the desire to "die with dignity," and yet not be aware of the real consequences of withdrawing or withholding various measures in regard to suffering. The physician should form a treatment plan with the patient, which should include plans to alleviate suffering. Such a plan might include the heavy use of analgesics. Many physicians are still reluctant to supply such patients, or any dying patients, with large doses of narcotics. There appear to be a few inappropriate sources of that reluctance.

Part of that reluctance may be due to a mistaken belief that such dosages are illegal. Part may be due to an irrational belief that the patient will become addicted, and if, as the result of some miracle, the patient recovers, he will be left with an addiction. Part of the reluctance may come from a belief that people should endure suffering. A more reasonable source of reluctance is the belief that the administration of narcotics may hasten the patient's death and thus be thought of as active euthanasia. However, the general trend of thinking is that it is permissible to administer large doses if the *intent* is only to alleviate pain, even though the side effect is to shorten life. (We will discuss this issue further when we speak of active euthanasia.) The physician who has qualms about administering large and frequent doses of narcotics to such patients should carefully and thoroughly examine the sources of his reluctance.

As with DNR orders, patients should be given time and opportunity to think about various alternatives. In some instances, where the patient is not sure, it may be possible to suggest a trial period of life-sustaining care or a trial period of palliative care. Under those circumstances, both patient and physician can get a sense of the actual consequences of administering or withholding treatment for that individual patient. In doing so, the physician must be very careful about interpreting a patient's reactions. For example, a competent patient who lacks the ability to speak may vehemently reject the insertion of a nasogastric tube. That may be evidence that he wants no nutrition and prefers to be allowed to die. On the other hand, he may simply be rejecting the discomfort of the tube, and might be willing to accept nutrition and hydration administered directly into the stomach through an incision.

Nutrition and Hydration

The withholding or withdrawal of nutrition and hydration has been a matter of special ethical concern. Physicians are usually willing to consider the administering of artificial nutrition and hydration, and even feeding and water by mouth, as medical measures. However, many nurses and some chaplains have been taught to consider these not as medical treatments, but as measures that are part of the ordinary and basic care that is due to any person, patient or not. For that reason (among others which we will discuss when we deal with issues relating to incompetent patients), there is often opposition to the withdrawal or withholding of nutrition and hydration, particularly if the patient is still capable of taking food and water by mouth.

In the broad sense, a recognition of persons' autonomy and their concomitant right to self-determination could be seen as entailing that persons have the right to reject the imposition of food and water whether or not food and water are conceived of as "care" measures or as "medical" measures. The general trend at present is towards treating artificial nutrition and hydration as no different from other medical measures, such as respirators. However, there has been resistance to allowing patients to reject food and water if they are still able to take them by mouth.

LIVING WILLS AND PROXY DESIGNATIONS

The concept of the living will is becoming quite well known to the public. There are two senses in which the term is usually used: (1) as a written expression of his wishes about treatment if and when the writer of the will were to become incompetent; and (2) as a request about the disposition of body and organs in the event of death.

We shall reserve consideration of the second sense for our discussion of the allocation of scarce resources. We shall speak here of the first sense, and only in regard to living wills that express wishes about treatment in the event of irreversible incompetency and either terminal or very serious illness.[13]

The living will is usually a person's request that physicians will refrain from aggressive treatment if he, as a patient, were to become terminally or seriously ill and no longer in a condition to request that treatment be terminated. However, it is important to note that a living will could just as well be a request *for* aggressive treatment under those same circumstances.

Living wills do not yet have full legal standing in all states. However, even in states that do not officially recognize them, they can be offered as evidence in court cases where a request has been made that an incompetent patient's treatment be terminated, or that extraordinary means be withheld. Here the concept of the "substituted judgment" becomes relevant.[14] Courts in states that do not officially recognize living wills may still allow them to be used as evidence for making a substituted judgment. That is, they can serve as evidence in determining what the patient would have wished were he competent.

While the concept of the living will seems to be a good one in theory, it can present some problems in practice. First, there is a problem about the length of time they should be considered to be valid. We would probably want a periodic renewal of such wills. It might not be just to hold someone to wishes he expressed when he was 20 years old now that he is 70 years old. Second, there may be a problem about the specificity of conditions that should be required to be written into living wills. Would we want persons to specify exactly the conditions under which they would want care or the withdrawal of care?—"I would not want treatment if I am in coma, have no reaction to painful stimuli, but do have reflexes. However, even under those conditions I would want care if I were suffering from a terminal illness for which a recent edition of the *New England Journal* announced a possible cure which is now in clinical trials." It is hard to imagine a living will that could predict all the contingencies that will arise in the course of an illness. Attempting to avoid that problem by creating generalized living wills creates its own problems. Thus, a living will may state that the patient would reject life-sustaining measures if he is in a condition "such that there is no reasonable expectation of recovering or regaining a meaningful quality of life."[15] While we may agree that a patient who is competent can decide whether he considers his present "quality of life" to be "meaningful," it may be hard for others to make that judgment about a patient.

One possible way to avoid some of the difficulties inherent in living wills is to permit or even encourage patients to designate some person to serve as their proxy in the event that they become incompetent. Some states already permit patients to do that, allowing them to give someone a "durable power of attorney for health care decisions." Presumably, such a designated proxy would know the patient well enough to be able to make a substituted judgment for the patient if the patient becomes incompetent. The proxy concept also presents some problems. Many patients are understandably reluctant to make such difficult decisions. It is not unusual for them to ask their physician to decide for them. Such patients may (again, understandably) use the proxy concept as a way of shifting the responsibility for hard choices unto a proxy who himself may be reluctant to make those choices. The proxy may not want the responsibility, yet may be unwilling to reject the role because of feelings of guilt. Moreover, even a willing proxy may face problems trying to make a substituted judgment. When a proxy or surrogate has been designated, the concept of initiating trial periods of palliative or aggressive care may also be appropriate.

HOSPICE CARE

Hospice care is another option for terminally ill patients. Hospice care may be given in a freestanding facility or else may be connected with home care. Patients who have been diagnosed with a terminal illness may choose hospice care. In the best of cases, hospices provide complete and well-trained support teams including psychologists, social workers, and specially trained physicians and nurses. The care is palliative, though usually the patient may have the option of changing his mind and requesting transfer to an acute care facility for aggressive treatment. Emergency facilities are made readily available to the patient for crisis situations.

While humane in principle, there are possible problems with the hospice concept too. Hospice care can be less expensive than aggressive hospital care. There already have been suggestions that terminally ill patients have an obligation to choose hospice care in order to lower medical costs to society. Terminally ill patients often feel that they are a burden to family and treatment teams. We may begin to find that patients who would really prefer aggressive treatment will be coerced into choosing hospice care or, for that matter, coerced into choosing other forms of "voluntary" passive euthanasia. On another level, we may face a situation where society begins to shunt money away from research for treatment and towards hospice care. Nevertheless, the physician should be aware that hospice care may be the best option for some patients and is, in fact, to be considered as a treatment option when obtaining informed consent and discussion options in the course of treating a terminal illness.

VOLUNTARY PASSIVE EUTHANASIA FOR COMPETENT NONTERMINAL PATIENTS[16]

There are instances in which a patient is not terminally ill, needs life-sustaining measures, and refuses those measures because he wants to die.

Most often these cases occur when a patient has suffered an illness that will allow him to live for an extended period, but will leave him in a condition in which he believes his quality of life will be unacceptable regardless of what treatments are available. The victim of severe burns facing months of excruciating pain and inevitable disfigurement, the established athlete or professional dancer left quadraplegic, the creative artist or intellectual with profound and irreversible brain damage that will prevent him from continuing his career, and the patient in the earlier stages of Alzheimer's are examples of such cases. Very often, such patients want to refuse treatment, but also want to receive palliative care in the hospital until they die. That is, such persons often are unable to attend to their basic needs, and may want to die in the hospital where toilet needs, bathing, and analgesics will be available to them.

These cases raise a number of difficult issues. In the case of imminently terminally ill patients, both arguments for passive voluntary euthanasia seem applicable. That is, the terminally ill patient can refuse treatment on the grounds that he has the right to refuse treatment, and can also offer a reasonably convincing argument that no treatment is the best treatment because aggressive treatment would offer no reasonable possibility of benefit.

The patient who is not terminally ill can offer the same arguments, but may be more convincing with the first. Certainly he can claim that he has the right to refuse any treatment, though that still raises problems to which we shall speak. However, his claim that treatment would offer him no reasonable benefit may be less convincing. The physician may believe that such a patient's judgment is clouded by psychological reactions to his condition, and that he could be convinced that readjustment and a meaningful life is possible. More strongly, the physician may believe that the patient has a duty to continue living and to adjust to his circumstances.

Although there must always be a presumption of competency, it is not unreasonable to take into account that the patient who has recently confronted a serious illness may go through temporary periods of denial, anger, and despair. In certain instances, such as burn cases, there may be a belief that the patient is in such enormous suffering that he is unable to make a rational decision. In these circumstances, we are faced with some difficult issues about determining competency, as well as difficult issues about the imposition of values upon a patient. For example, should the patient be forced to go through a "cooling-off" period before he is allowed to refuse treatment? Should he be forced to go through a trial period of treatment before he is allowed to refuse treatment?

Suppose that, after a cooling-off period and counseling, he continues to insist on refusing treatment, and the institution or physician believes that he ought to have treatment. Should such patients be considered to be suicidal, and forcibly treated? Does the institution have the right to discharge him from the hospital?

The courts so far have been mixed in their reactions to these cases. Sometimes they have found for forced treatment; sometimes they have ruled that the hospital has the right to discharge such patients. In other instances, they have allowed the patient to refuse treatment and, at the same time, forced hospitals either to provide palliative care or to arrange for a transfer to a facility that would accede to the patient's wishes.

As for the obligations of the treating physician, the trend has been towards permitting physicians to withdraw from treatment of patients if they believe they cannot morally aquiesce to the patient's desires; however, the physician has also been given the duty to arrange a transfer of the case to a physician or facility that will accede to the patient's wishes.

NONVOLUNTARY PASSIVE EUTHANASIA

Aspects of nonvoluntary euthanasia raise some of the most difficult issues in medical ethics. We may subdivide the areas where issues arise into (1) euthanasia involving ill adults who are incompetent as the result of illness; (2) euthanasia involving adults incompetent as the result of a developmental disability; and (3) euthanasia involving neonates.

INCOMPETENTS

Terminally Ill Adults

Here we shall speak of adults who have become irreversibly incompetent during the course of a terminal illness. Typically, such persons have not made express wishes about their treatment preferences before becoming incompetent, or chosen surrogates or proxies. These cases are very common. We should point out again that some of the difficulties that will emerge in this section could be avoided if patients were told the truth about their prognoses, and if persons were better educated about their options before falling ill or becoming incompetent. These sorts of cases, while common, still present many ethical difficulties and there is not yet any ethical or legal consensus that fully answers the difficulties. The following includes a synopsis of the current thinking on this issue and may give the reader an idea of what laws will require once they have been set into place. The reader is also referred to Chapter 2, "Informed Consent and the Right to Refuse Treatment."

A typical case involves an elderly patient confined to a nursing home or other chronic care facility. The patient is very demented and suffering from a number of complications. His condition has worsened, and he is now on a respirator and receiving tube feedings. He has gone into cardiac arrest a number of times. He has never made explicit requests about treatment. The question arises among staff as to whether a "no code" ought to be called and, perhaps, questions about continuing tube feeding and treating infections.

First, we reiterate the importance of discussion with a patient of his prognosis and alternatives in regard to treatment *before* he becomes incompetent.

If the patient is irreversibly incompetent, then a "substituted judgment," as has earlier been described, must be made. A court or surrogate must try to decide what the patient would have wished were he competent. If that cannot be determined, then the decision must be made on the basis of what is thought to be in the best interests of the patient. Questions about medical costs, provider workloads, and scarcity of technical resources will be allowed to have only a *tertiary* impact on such decisions.

Here again we see the primacy of autonomy theory approaches to decision making. As a person, I might believe that if I were in the position of the patient, I would want no treatment or I would want aggressive treatment. However, as a surrogate, my role is not to impose *my* preferences upon the patient but, instead, to try to give voice what *his* wishes might have been.

As with informed consent in general medical situations, hospitals or attending physicians will often turn to the next of kin to make decisions when the patient is incompetent. We should point out that as far as the law is concerned, it is by no means universally accepted that the next of kin is automatically empowered to make decisions for the patient, although many courts will uphold such decisions. There has been some movement towards automatic acceptance of a next of kin's decision; but the reader should be warned that if the law does automatically accept a next of kin's decision, there will remain serious ethical issues. Family members are not necessarily in the best position to make a judgment about the wishes of the patient. Even the next of kin may not know the patient well. Moreover, family members who are close to the patient are usually distraught and under a great deal of stress. They may have a mixture of feelings—anger, guilt, sadness, denial—which may prevent them from making a reasonable judgment. Too, sometimes family members will disagree among themselves about whether treatment should be terminated. Sometimes a friend, or a member of the health care team, or a clergyman may know the patient's wishes better than his family does. Some would say that the decision ought to be put into their hands.

It should be remembered that the role of any surrogate is to try to determine what the patient would have wanted. The physician should explain the options to the surrogate, but not make the choice. When there is time, the surrogate ought to be given the opportunity to make an unrushed judgment. If the surrogate decides to terminate life-sustaining measures, then he ought to provide a signed and witnessed statement, and full notes ought to be written into the case record. The concept of "trial periods" of treatment or palliative care may be applicable here. Nevertheless, even they pose great conceptual difficulties. The underlying criterion often resorted to for withholding or initiating treatment (including nutrition and hydration) is whether the measures will cause or alleviate suffering for the patient. That is not an easy determination to make for very ill or very demented patients. We do not really know if the very demented patient suffers from feelings of hunger and thirst if he is denied food or water. Reports from competent persons who have gone on hunger strikes indicate that a point is reached at which feelings of hunger or thirst are no longer present. But the evidence is only anecdotal. Again, we do not know if the somewhat demented patient who rejects the nasogastric feeding tube wishes to die, or simply does not wish the insertion of the tube. We are not sure if the demented patient with a fever is suffering. It would seem that patients in deep coma or in persistent vegetative states are beyond feelings of pleasure or pain, but that conclusion is somewhat controversial.[17] There is a body of literature available suggesting ways of alleviating any possible suffering in such patients,[18] but the underlying questions remain.

If close family and friends are not available, or there is any question about a surrogate's capability of making a decision, the attending should make an effort to speak to those who were in close contact with the patient. This would include other members of the treatment team, including nursing staff, social workers, and pastors. Depending on the hospital, an ethics committee may be available to help deal with particularly problematic cases.[19]

In certain instances, the physician may have an obligation to contact the hospital administration or attorney to protest or question the decision of a surrogate. Aside from written documents (e.g., living wills), reports of verbal wishes made during the patient's periods of competence may also be offered as evidence in forming a substituted judgment. Thus, for example, Karen Ann Quinlan[20] had expressed a wish during her lifetime not to have extraordinary means used if she should ever become comatose. While reports of her wishes were not directly used as evidence in the court decision to withdraw a respirator, they were certainly a major factor in her parents' decision to request that she be allowed to die.

In the Brother Fox case,[21] evidence was offered that Brother Fox had lectured on patients' rights to have extraordinary means withdrawn, and had often expressed a wish not to have such means used with him. This evidence was used by the court in making a determination that treatment be withdrawn when he fell into coma.

Specifying precisely what a person's wishes were is often difficult, since most persons still do not think or talk in depth about such possibilities before they are ill. In many instances, forming a substituted judgment becomes quite difficult. The fact that the use of extraordinary means has become so "ordinary" is an important reason for persons to consider their desires about treatment before they are unable to make a decision. It is also an important factor in truth-telling to patients. The patient who doesn't know the truth about his prognosis cannot really express wishes about his desire for or against heroic measures. The problems may diminish somewhat since the public is becoming more aware of the options available. Persons are beginning to express wishes about their future treatment.

The decision to terminate treatment should not be taken lightly, if only for the fact that the practice has the potential to lead to abuse of the sorts described above in the section on voluntary passive euthanasia.

Retarded and Psychotic Patients

We have lumped these two different groups together because in both patient groups there are often long periods of incompetency. Linked to this there tends to be confusion about their rights and our obligations in regard to their medical treatment.

First, it must be stressed that a finding of retardation or psychosis is not the same as a finding of incompetence to make medical decisions.[22] If a question of termination of treatment comes up in the course of treatment of either a psychotic or retarded patient, there will have to be an evaluation of the patient's competency in regard to treatment that is separate and apart from the diagnosis of retardation or psychosis. Once that determination has been made, much of what we have said about incompetent "normal" adults is applicable to these special cases.[23] However, with very retarded patients the problem of making substituted judgments becomes difficult. There may be no reasonable means or evidence for determining what their wishes would have been were they competent, since it is possible or likely that they never were competent. With very psychotic patients, we face the difficulty of determining that their incompetence is in fact irreversible. That is, irreversible incompetence or incapacity may be reasonably ascribed when there is direct evidence of organic brain damage, such as in patients with advanced Alzheimer's, or patients in deep coma. The same certainty may not be possible with nonorganic psychoses. Furthermore, if the psychosis has been longstanding, we face the same problems about forming substituted judgments that we do with the severely retarded.

Attempted solutions may be either to form a substituted judgment based on what treatment an "average reasonable person" would be likely to accept, or to make a judgment based on the "best interests" or "quality of life" of the patient. Both proposed solutions have their problems. The idea of the "average reasonable person" is a legal concept decided in each case on the basis of historical precedents and upon jury decisions. It is questionable how applicable the concept is to these rather new types of cases. The concepts of "best interests" and "quality of life" are often referred to in writings on these issues. Here too, we face difficulties. The "quality of life" I would be willing to endure may differ from the quality of life another person might be willing to endure.[24]

In some cases, it may be comparatively easy to apply the concept of "best interests." If, for example, resuscitation will leave the terminally ill patient still in overwhelming uncontrollable pain, then it may be possible to say that resuscitation is not in the best interests (or to the benefit) of the patient. However, other cases may be more difficult. Suppose, for example, the very retarded terminally ill patient faces resuscitation or other procedures that may cause him pain and also distress and anxiety because of his inability to comprehend what is going on. Should that distress and anxiety weigh in favor of not initiating or continuing life-prolonging procedures?

Defective Neonates

The problem of whether or not to treat defective neonates is one of the most difficult faced by the physician. Once again we have an issue that has been magnified by recent technological developments. In the past, the majority of fetuses with severe genetic defects were spontaneously aborted.[25] Many of those defective neonates[26] that did come to term could not be saved and died shortly after birth even with the most advanced treatment available. In practice, it was not unusual for physicians to "informally" terminate treatment of defective newborns in order to let them die more quickly.

We now have technology available that permits prenatal diagnosis for defects, drugs available to prevent spontaneous abortions, and techniques that allow us to keep many severely defective newborns alive for at least a short time and allow us to correct at least some of their defects.

The present political and social climate has brought a concerted effort to place strictures upon abortion and, concurrently, a related effort to force physicians to use all means available to save all newborns. At the time this is written, there has been no legislation that directly forces treatment of all newborns. However, there have been indirect pressures brought to bear that tend to have the same effect. The first attempt to force treatment was an effort to put termination of treatment under the rubric of "child abuse." Hospitals were informed that federal funds would be withdrawn from any hospital that didn't treat all newborns. At the same time, the government distributed posters to all hospitals to be placed in pediatric intensive care units. The posters contained a warning about the termination of federal funding, and a 24-hour hotline number to which reports of "child abuse" could be reported. The directive was overthrown by the courts.[27] It was replaced by a directive stating the hospitals that refused to use all means possible to save all defective newborns would be liable for charges of discrimination against the handicapped.

This is where things now stand, and perhaps it is a good place to initiate discussion of the issues. Is, in fact, a refusal to treat defective newborns always an instance of discrimination against the handicapped? Many other issues relating to the care of defective newborns are of secondary importance. For example, considerations of the drain on the emotional and financial resources of the family are emotionally compelling, but must remain secondary to consideration of the best interests of the patient, whether the patient is an adult or a newborn. Similarly, questions about the drain on society's and the hospital's resources must also remain secondary.[28] The utilitarian benefits of learning how to save future patients by trying methods of aggressive treatment of present hopeless cases are also enormous. However, they too should probably take second place to considerations of the good of the particular patient actually present.

The premise underlying the claim that refusal to treat defective neonates is discriminatory is that a refusal to treat is always a harm to the patient. Yet, as we know from consideration of the treatment of adults, this premise is a questionable one. It is not always certain that refusal to use aggressive measures is always harmful to the patient or is always synonymous with neglecting to treat in the best way possible.

For example, with the anencephalic infant, as with the adult in a persistent vegetative state, it is doubtful whether it makes sense to say that the patient can suffer significant "harm" or be significantly benefitted by any treatment. There are no conceivable measures that will bring such infants to any semblance of personhood.[29] Depending on the degree of anencephaly, there may be moral issues about pain management. That is, some of these infants may be able to feel pain, and there may be a duty to prevent or relieve such pain, although no duty to try to keep them alive.

In instances in which the infant has a multitude of problems—some correctable, some not—we are faced with a bifurcation. (1) There are those cases in which the most intense "treatment" can significantly prolong the infant's life, but leave the infant in a condition in which its capability for mental function will always be very limited or always overcome by pain or other disabilities. Here it may be argued that it would be discriminatory to subject the infant to "treatment." That is, in these cases, the best treatment may, in fact, be no treatment. (2) There are those cases in which any procedure might repair some defects, but have the effect of instilling pain without significantly prolonging the infant's life. For example, an infant may be born with a multitude of serious defects, only some of which are treatable. Here it might also be argued that the best treatment is no treatment.

With defective neonates, we are faced with very difficult tasks both of forming a substituted judgment and of deciding about benefit to the patient. It is not clear if we can even apply the concept of "substituted judgment" unless we think of it in terms of "average reasonable person."[30] Thus, decisions may have to be made on a "best interests" principle. As with incompetent adults, there are problems with assuming that the next of kin are automatically entitled to serve as surrogates. In the case of defective newborns, the next of kin are, of course, almost always the parents of the infant. While parents are usually given a legal right to decide about medical treatment for their children, courts may override a parent's decision if the court believes the parent's decision is not in the best interests of the child. Given that parents do have a prima facie claim to make a decision, then the requirements of informed consent in general are applicable here. Parents of the child should be given the prognosis and told the options for treatment. The situation and options should be very fully explained to them. Immediate difficulties that might arise should be apparent. The parents will probably be distraught and have feelings of guilt, denial, and perhaps anger. Parents may also disagree about the best course of action. Physicians may disagree with the decision of the parents. As with terminally ill adults and their families, liaison providers should be available to speak to the parents. Any discussion with the parents should include the point that they must consider the best interests of the infant as primary to their decision. For example, those parents who may believe that taking care of a severely defective child can be a fulfilling experience for them should be made aware that that is really a consideration of their own interests, not of the infant's.

The hospital may have a special neonate committee empowered to deal with these issues. However, it should be apparent that any such committee will face the same problems.

There seem to be very good arguments to support withdrawing treatment for infants with severe defects such as anencephaly, and for infants who will have a poor prognosis no matter what is done. However, the major ethical problems arise when dealing with infants who are "borderline." How defective must an infant be before withdrawal or withholding of treatment is justified or called for, and before a judgment could be made that the prognosis is for an unacceptable quality of life? Suppose, for example, the infant is mildly retarded with a duodenal atresia. May the parents request that no corrective measures be taken, and the infant be allowed to die? Or suppose the infant is diagnosed with a serious genetically linked and incurable disease, such as Huntington's chorea, that will not emerge until late childhood or adulthood?

Moreover, while we spoke of "significantly" extending the infant's life, it is not a simple matter to pinpoint what a "significant" extension is. Is it 6 weeks?—6 months?—a year?

The dilemma seems apparent. On the one side, there is the belief that some infants simply have a prognosis that will leave them better off dead, and thus better untreated. On the other side, there is the belief that an infant's life should not be ended for "trivial" reasons. For example, many would believe that parents dissatisfied with the sex of an infant with fatal but easily correctable defects should not be allowed to request no treatment simply because they prefer a child of the other sex.

It would seem that the issues relating to those infants who will die no matter what is done are directly analogous to issues having to do with the incompetent terminally ill adult. Similarly, the issues relating to the infant who is anencephalic are analogous to the issues relating to the adult in a persistent vegetative state.[31]

As is often the case in medical ethics, issues overlap. We refer the reader to the sections on abortion, the right to refuse treatment, and the section on genetic engineering for a fuller discussion of the issues we have touched on here.

VOLUNTARY ACTIVE EUTHANASIA

At this writing there is an absolute prohibition of active euthanasia in the United States. For example, the American Medical Association expressly forbids active euthanasia.[32] The practice is also forbidden by law.

There are groups in the United States that have organized to lobby for legislation permitting active euthanasia.[33]

In Europe there appears to be some recent movement towards the legalization of active euthanasia. In a case in the Netherlands,[34] a high court requested that a study be made to set guidelines for active euthanasia.

While the legal tradition in the United States is strong in its prohibition of active euthanasia,[35] there are compelling ethical arguments in favor of the practice. It might be argued that if we allow passive euthanasia in cases in which we believe that continued life would only prolong suffering, why not take the extra step and alleviate all suffering in these cases? A terminally ill patient, if conscious, will suffer if a respirator is removed. Regardless of our ability to manage pain, many terminally ill patients will go through enormous anguish. It would seem more humane to administer a lethal overdose of narcotics or other drugs to such patients if they were to request it. There are those who say that such patients should not be given overdoses, but may have life-sustaining means withdrawn and, when possible, be given sufficient analgesic to keep the patient insensate. But, in counterargument, it could be said that if such analgesic dosage has the effect of rendering the person insensate until he dies, and has the side effect of shortening life, then there is little difference between keeping the patient unconscious until his death, and painlessly killing him.

But there are arguments against active euthanasia that are not easily answered. First, there is a utilitarian argument that allowing active euthanasia will set a dangerous precedent and will eventually lead to involuntary active euthanasia. Second, there is a widespread belief that physicians simply are not in the profession of causing death. Third, there is the argument that passive euthanasia is really an instance of a patient's right to refuse treatment. It is a rejection of intervention, while active euthanasia is a *request* for intervention.

Each argument has merit. The first, utilitarian argument points out that permitting voluntary active euthanasia could, in fact, set a dangerous precedent. However, that may only serve as an argument for extreme caution in setting guidelines for active euthanasia. Such guidelines would have to include strict provisions for the determination of the *competency* of the patient and the *voluntariness* of his request for active euthanasia.

The argument that physicians should not be in the business of causing death sometimes rests on an implicit assumption that good treatment always means aggressive treatment. If, in fact, we agree that a patient sometimes may be better off dead and that no treatment can be the best treatment, it seems inevitable to conclude that the actual provision of death may sometimes count as a preferred treatment modality. The claim that the physician has the duty to use aggressive treatment because there is always hope of a remission or of new developments, seems to be an empty one. There are cases in which nothing foreseeable in the near future will help the patient.

However, the second argument may take a different form; the arguer may claim that the physician's task is always to use aggressive treatment, but that his preference may be overridden by the patient's right to refuse treatment. That is, that the second and third argument are the same. Following from this, it would be argued that the patient has the right to refuse life-sustaining means, and thus a right to passive euthanasia. However, the argument might continue, that doesn't give the patient a right to active euthanasia. Thus, the physician cannot be forced to administer active euthanasia, even if the patient requests it. While that may be a convincing argument against *requiring* physicians to use active euthanasia, it will not serve as an argument against *permitting* a physician to use active euthanasia if he has the patient's consent and if the physician is willing.

Moreover, the argument doesn't really answer the claim that in some cases a quick and painless death may be the best way to benefit the patient. To sort out the problem of active euthanasia is a task that likely will involve us for a long time to come.

Chapter 5

Psychiatric Ethics

A woman, about 40 years old, is observed living in a refrigerator packing case placed over a heated grate on a street in a large city. She has lived there for about 2 years. She is filthy, and is sometimes seen defecating and urinating on the street. She sometimes curses passersby. She sometimes accepts handouts, sometimes burns paper money given to her. It is midwinter, and temperatures are below freezing. Approached by city social workers, she refuses to enter a city shelter, claiming that the shelters are dangerous and that she prefers the streets. She claims that she manages very well the way she is. Approached by psychiatrists on a number of occasions, and asked to enter a hospital, she is abusive and refuses to speak with them. She is forcibly brought to a psychiatric hospital and involuntarily committed under a law that says such hospitalization is permissible if psychiatrists judge that a person is either a danger to himself/herself or a danger to others, or unable to appreciate that he/she needs psychiatric or medical treatment.

Many of the ethical issues that come up in medical care are directly tied to issues of mental health. When we wonder whether a patient should be given information about his poor prognosis, we are probably really concerned with the underlying question "Can he take the news, or will he become too depressed?" If we wonder about the feasibility of getting truly informed consent from patients or research subjects, our concern is often about the ability of the patient or research subject to both understand what is at stake, and make rational decisions. If we question a patient's decision to refuse treatment that we think is appropriate, we are often really questioning the rationality of his choice.

There are also a number of ethical and conceptual issues that are peculiar to mental health care.[1]

To begin with, the psychiatrist's patient population is by definition less than fully autonomous. Patients come to treatment or are brought to treatment for the very reason that their abilities to control their own lives and decisions are impaired. Thus, those who stress the importance of recognizing patient autonomy at the onset of medical treatment face an initial dilemma in dealing with the psychiatrist–patient relationship. Moreover, while the conceptual issues involved in giving value-free definitions of "health" and "illness" have by no means been totally resolved in the area of somatic medicine, in practice they relatively infrequently present problems. In discussing mental health care, however, the problem of defining "mental health" and "mental illness" sits up front and is always a real presence in any discussion of ethical issues in the actual practice of psychiatry. Tied to this conceptual problem are important conceptual disputes about both the etiology and treatment of mental disorders.

In some areas of psychiatric "care"—involuntary

commitment, and much of the practice of forensic psychiatry, to name two—we are not even sure that health care is involved in what the psychiatrist is asked to do. What has been called the ethical issue of "double agenthood" is especially cogent here. More than any other type of physician, psychiatrists are often asked to play two seemingly conflicting roles. On the one hand, they are asked to protect the interests and desires of their patients and, on the other hand, they are asked to protect the public or other individuals from the actions of their patients.

CONCEPTUAL ISSUES: "HEALTH" AND "ILLNESS"

While it is often said that medicine is more of an art than a science, the claim is really metaphorical. The fact is that medicine is more science than art both in its goals and methodology. It may still be true that skills in both treatment and diagnosis cannot be totally taught by giving physicians "checklists," and it may also be true that medical expertise still requires "hands-on" learning and experience. Nevertheless, the fundamental approaches to research into the causes and cures of diseases, and the fundamental approaches to the diagnosis and treatment of diseases, follow the methods of scientific explanation and testing. The clinical "picture" of a patient is drawn with facts, not with paintbrushes.

Every science strives to find value-free definitions of its crucial concepts. For example, we would not expect the definition of "chlorine" to vary from culture to culture, or be dependent on particular social norms or upon the ethical or religious values of individuals. Certainly, the concepts of "mental health" and "mental illness"[2] are crucial to psychiatric care. Unfortunately, these concepts have had a past history that did tie them to value-laden preconceptions. That is not surprising, since psychiatry deals with human behavior and it is easy to unanalytically categorize behavior or character traits that we find unacceptable as "abnormal."

Mental Illness as Abnormal Behavior

One example of a value-laden approach to defining mental health and mental illness is defining the concepts in terms of societal norms of what counts as acceptable and unacceptable behaviors. This is a claim that any behavior pattern that significantly deviates from general norms of behavior is a sufficient sign of the presence of mental illness. This claim itself may mean two very different things, depending upon how "normal" is defined.

First, "normal" may be defined in a statistical way. That is, normal behavior is defined as what most people do most of the time; abnormal behavior is defined as behavior that deviates from the majority's behavior. So, for example, most people in our society who have the financial means or opportunity to do so will live in homes rather than on the streets. Under this definition of "normal," there would be a prima facie assumption that anyone with the means or opportunity to live in a home or apartment or shelter who "chooses" to live in the streets instead is mentally ill. Or, for another example, most people in our society do not commit crimes of violence against others. Thus, there is sometimes a prima facie assumption made that anyone who commits crimes of violence is mentally ill.

Second (and, as we shall see, not entirely separable from the first), "normal behavior" may be defined as behavior that is consistent with some conception about what should be "natural" behavior for human beings qua human beings. This is the type of conception of "normal" inherent in natural law theory,[3] for example. As we said in Chapter 1, conceptions of normalcy according to natural law theory have had an enormous influence on psychiatric conceptions of what counts as mental illness.

In a rough sense, definitions of "abnormal" behavior derived solely from statistical deviancy fit very well with relativistic[4] conceptions of morality. "Essential" definitions of normalcy, such as natural law definitions, are absolutist. One implication of essential definitions that goes contrary to the implications of statistical definitions is that large segments of a society (or even whole societies) can be "mentally ill" if the general behavior of the population deviates from what the theory regards as required normal behavior.

For example, a "statistical" approach might classify homosexual behavior as within the bounds of mentally healthy behavior if homosexuality were widely practiced in society. An "essential" normalcy theory, such as the Catholic interpretation of natural law theory, believing homosexuality to be contrary to the purposes of human essence, might classify such a society as having a high incidence of a mental disturbance.

The two approaches are sometimes mixed in some people's thinking. Thus, those who claim that violent behavior is an indication of mental disturbance may base that claim upon a perception that such behavior is statistically deviant, but they may also claim that violent behavior is inconsistent with basic human nature.

Abnormal as Deviant Behavior

Objections to statistical deviancy theories as appropriate ways of defining mental disturbances have come from a few major fronts. There have been objections on the ground that the majoritarian approach is likely to lead to the unjust labeling of various groups or individuals as "mentally ill" simply because they are nonconformist, be they politically different, or sexually different, or lead unusual life-styles. "Radical libertarians" such as Dr. Thomas Szasz[5] may make a strong claim that, unlike the criteria used in diagnosing somatic diseases, there are no value-free criteria for diagnoses of mental illness. They may believe that any purported diagnosis of mental illness is simply a reflection of what a society considers unacceptable behavior, and thus such a diagnosis is "unscientific." In this view, it is not only unscientific but also morally unjust to apply the label "mentally ill" to *any* person, because there is always a pejorative element to the categorization. This pejorative element can be seen in cases in which politically deviant behavior is taken as indicative of mental disturbance. The label denigrates the political dissenter by implying that his ideas are the ravings of a madman and not to be taken seriously.

The radical libertarians may extend their charge even to instances in which deviant behavior is traceable to organic brain "dysfunction." They may argue that what we count as a "dysfunctional" brain is determined by what we count as the "disturbed" behavior caused by that "dysfunction," and calling that behavior "disturbed" is, in turn, a value-laden description. For example, Tourette's syndrome, in which the patient has inappropriate outbursts of speech, sometimes obscene, may be traceable to particular lesions in the brain. The radical viewer of mental illness may say that the very categorization of such obscene outbursts as "inappropriate" and as indicators of a mental disturbance is itself dependent on societal values.

By extension, some of these theorists (and we shall speak more about this when we deal with the issue of involuntary commitment) believe that forcible psychiatric treatment or institutionalization is a violation of a person's civil liberties.

Sometimes a variation of this sort of view appears in the form of a cultural relativist approach. Some may claim that what is labeled mentally disturbed behavior in one culture may be considered quite normal or even admirable behavior in other cultures, and go on to a further claim that it therefore follows that there are no universally applicable and absolute definitions of mental illness. These objectors also believe that it is wrong to categorize persons as "mentally ill" and especially wrong to believe that in doing so we have found the equivalent to the value-free discovery of the presence of chlorine in a water sample.

Other objectors to the statistical deviancy view agree that statistical deviancy alone is not a sufficient criterion for the ascription of mental illness, but hold (unlike the libertarian-relativist) that there may be some other supportable criteria that are absolute and not value laden.

The cultural-relativist claim that what counts as mental illness is totally culture-bound is open to question. There is evidence that most cultures do, in fact, categorize some persons as mentally ill, and do distinguish between persons who are "delusional" and persons who are "seers" or "prophets." More profoundly, the claim is open to the same objections that we made in Chapter 1 to cultural-relativistic claims about ethical values: Even if there are cultures that classify persons whom we call mentally disturbed as great prophets, that alone doesn't prove that there are no independent and universal criteria for the ascription of mental disturbance, or that some of the culture's ascriptions are correct. As we noted there, the fact that a culture believes that epilepsy is a punishment for sin and can only be cured by prayer, atonement, or God's grace doesn't prove that our medical theories about the etiology and cure of epilepsy are value-laden and not better explanations.

The more general objection that, unlike physical illness, mental illness is *only* definable in terms of what a society considers to be normal and abnormal behavior raises deeper questions.

In reply to this objection, it might be pointed out that our concepts and nosologies of physical health and illnesses might be open to the same sort of charge. One might question whether our definitions of, and criteria for, physical illness are always independent of what society considers normal and abnormal states of being. For example, at what point do we justifiably say that someone is "too short," "suffers from pituitary dwarfism," and thus "should" have treatment?[6]

Moreover, while it might be reasonable to say that any definition of health or illness that is solely dependent on statistical norms is faulty, that doesn't prove that the use of statistical norms is the only possible way to approach definitions.

Abnormal as Behavior That Deviates from Essence

Accepting that mental health/mental illness can have absolute and value-free definitions really commits one to accept some sort of essentialist claim. That is, regardless of how the members of society actually do behave, there are some ways in which, as human beings, they *ought* to behave. And, depending on the reasons persons deviate from that acceptable behavior, we are justified in diagnosing some of them as mentally ill. However, there are various essentialist claims about what constitutes mentally healthy behavior. Some of these claims are at variance with one another.

Philosophical Theories and Mental Health

In Chapter 1 we mentioned that each of the ethical theories is also a theory about mental health. Each theory tries to give an account of moral good and of right actions. In those senses, at least, they are speaking about the ways persons ideally ought to behave and that is at least somewhat congruent with the concept of mentally healthy behavior. As theories of philosophy rather than theories of psychology or psychological pathology, they tend to present "bare-bones" descriptions of, or only implications about, mental health. The theoretical connection between ethical behavior and mentally healthy behavior goes back at least as far as Plato, who claimed that the two are synonymous.[7]

Natural Law Theory

As we have said, the essentialist natural law theory has had the most influence on past and recent conceptions of mental health and mental illness. The theory, at least as presented by thinkers like St. Thomas Aquinas, is quite specific in delineating what types of behavior are in accord with our essence as human beings and thus categorized as "rational." We have already mentioned some of the types of behavior that natural law theorists have considered to be contrary to our nature as human beings: sexual behavior that does not have the intent of procreation and is outside the context of marriage. This includes masturbation, ejaculation outside of the vagina, homosexuality, the use of birth control devices, sex outside marriage, and bestiality.

Natural law theory sometimes has difficulty distinguishing between "unacceptable" behavior that is caused by medical pathologies and thus is nonvoluntary, and so categorized as "illness," and unacceptable behavior performed voluntarily, and thus blameworthy. Thus, for example, those instances of "unnatural" sexual behaviors mentioned above are sometimes thought of as morally reprehensible, sometimes as "sick," and sometimes thought of as a mixture of both reprehensible and sick—as "moral disorders." When the behavior was considered to be nonvoluntary, then those who exhibited that behavior were thought of as mentally disturbed. If the behavior is voluntary, someone who exhibits that behavior is morally blameworthy.

Objections to Natural Law Definitions

Natural law definitions of mental disorders have been falling out of favor for a number of interconnected reasons. First, the theory itself has many teleological elements that are unacceptable to modern science. Early formulators of the theory assumed that the universe and everything in it, including persons, were designed with a purpose. The idea of a purposive teleological universe has been discarded by the modern sciences and has been replaced with explanations that still speak of laws of nature, but exclude any talk of a purposive design of those laws.[8] In biology and medicine, talk of "purposes" has been replaced by talk of "functions."

Second, the development of more and different "scientific" theories of psychology have thrown some of natural law theory's explicit claims about "normal" behavior and its explicit claims about "human nature" into doubt. For example, Freudian theory presents a different picture of human nature. Also, reexamination of the justifications for natural law claims, many of which are still based on Aristotle's ancient accounts, has opened them to charges of bad science and cultural bias.

Third, competing ethical theories of "good," and of "right action," and of the "essence" of humanhood began to influence intellectual thought.

Utilitarian Theory

One such competing theory is utilitarianism.

As we said in Chapter 1, utilitarian theory has much in common with behavioristic psychology. To oversimplify both theories, both claim that our essential drives are simply for the pursuit of pleasure and the avoidance of pain. In that sense, we can say that the mentally healthy person is one who successfully satisfies these drives, or, at least, has the internal capability to satisfy those drives, if not frustrated by forces outside himself. The desirable state of being for humans is one in which their desires are fulfilled over a long period, and for both utilitarian and behavioral psychology theories, our desires are always ultimately analyzable as desires for pleasure and desires for the absence of pain. Clearly, the types of activities that produce pleasure for humans are quite varied and complex as opposed to, for example, the activities that might produce pleasure for a planaria.

In classical utilitarian theory, at least, there is no attempt to give a specific list of desirable (mentally healthy) behaviors. Along with that, there is a rejection of classical natural law criteria for desirable behaviors. In fact, John Stuart Mill, a founder of utilitarianism, was publicly active in criticizing natural law-based claims such as that women who chose careers instead of marriage and children, homosexuals, and other social "deviants" or "eccentrics" were necessarily to be thought of as mentally ill or immoral.

Social Contract Theory

Social contract theories also tend to theorize about human essence and thus at least imply theories of mental health and mental illness. We presented one such theory, that of Hobbes, in Chapter 1. His view was that humans are selfish by nature and thus have natural tendencies to get whatever they can to satisfy their own needs and desires. Because of competition from others who have similar needs and desires, only some of an individual's interests can be satisfied because he must "bargain" with other individuals. In order to bargain with others, an individual needs at least the intellectual capability to understand contracts, and also the ability to control his behavior so that he can conform to the requirements of the contracts he has with other individuals and with the government that protects him. Unlike natural law theory, there is no claim in Hobbes that we have any *natural* desires for love, community with others, or empathy for others. Therefore, these are not considered essential to mental health.

We have not spoken of other and contrasting versions of social contract theory. For example, Rousseau,[9] a social contract theorist, believed, contrary to Hobbes, that human nature included innate instincts, such as affection for others and a sense of community.

Kantian/Autonomy Theory

Kantian theory tends to discard the claim that there are specific instinctual drives that humans must satisfy in order to be "mentally healthy." For modern versions of Kantian theory, the only important essence of human nature is our possession of a "rational will"—our capacity to be autonomous. We have spoken extensively about that ability in Chapter 1. The ability includes an intellectual capability to comprehend both the reasons for our possible decisions and the probable consequences of our decisions. Those abilities, in turn, necessitate having a capability to examine facts, and draw reasonable conclusions from those facts. More specifically, and contrary to some theories of the human psyche,[10] the theory presumes that we are not entirely bound and determined by instinctual drives, or early conditioning. That is, we have "free will." Possession of a rational will and the ability to exercise autonomy are the characteristics of the mentally healthy person. While accepting that we have innate drives for pleasure and freedom from pain, happiness or pleasure or freedom from pain are not considered essential to mental health. In Kant's view, at least, a desirable and moral action may cause unhappiness for the actor. Conversely, one can be happy and yet not mentally healthy. For example, we can imagine that a person can achieve happiness by being passive in life. After all, being responsible for one's decisions in life can provoke anxiety, and it might be easier to allow oneself to be "infantilized" and have someone else make decisions. But to be passive in life is inconsistent with mental health.

Kant himself gave little thought about the ways in which persons can pathologically lack the ability to exercise a rational will, and gave little thought about the causes for such pathologies.[11]

Theories of Psychology

This is not a text in psychological theory and so we cannot deeply examine the various theories about the causes of mental disturbances. There are a number of competing theories (e.g., psychoanalytic, cognitive, social, organic), each with its particular implications for the nosology of mental disturbances, etiologies, and treatments. Proponents of each theory also have internal disagreements. The disputes among the proponents of each model have occurred on methodological grounds, with charges questioning the validity of evidence supporting each theory. Disputes have also arisen on ethical grounds, with questions about the possible existence of implicit and unsupported value biases in the theories. For example, there have been charges that classical Freudian theory contains unsupported and biased claims about the "essences" of women.

These disputes are important factors in the present reevaluation of the definitions of "mental health" and "mental illness."

The general trend towards acceptance of autonomy theory as the fundamental approach to the physician–patient relationship is being paralleled by a similar trend in the attempt to find value-free definitions of "mental health" and "mental illness."[12] There seems to be a movement away from classifying behavior that deviates from statistical norms, or deviates from natural law criteria, as prima facie indications of the presence of mental illness.[13] The movement seems to be toward defining mental health and mental illness in terms of autonomy. The goal of treatment is seen as an attempt to restore the patient's autonomy, rather than as an attempt to make the patient conform to any specific behavior pattern. In addition, there has been an attempt to shy away from strict adherence to the claims and implications of any one model of psychology.

Examples of these shifts can be seen in the approaches to etiology and classifications of mental disturbance found in the *Diagnostic and Statistical Manual* (DSM-III-R) issued by the American Psychiatric Association. There is an attempt to incorporate the implications of various explanatory models of psychology in the manual.

In addition, the general definition of "mental disturbance" given in the manual reflects a change towards accepting the "lean" definition of mental disturbances as constraints on a person's internal autonomy:

> In DSM-III-R each of the mental disorders is conceptualized as a clinically significant behavioral or psychological syndrome or pattern that occurs in a person and that is associated with present distress (a painful symptom) or disability (impairment in one or more important areas of functioning) or with a significantly increased risk of suffering death, pain, disability, or an important loss of freedom. . . . Whatever its original cause, it must currently be considered a manifestation of a behavioral, psychological, or biological dysfunction in the person. Neither deviant behavior, e.g., political, religious, or sexual, nor conflicts that are primarily between the individual and society are mental disorders unless the deviance or conflict is a symptom of a dysfunction in the person as described above.[14]

Here we see a partial acceptance of a view of mental disturbances as particular limitations on a person's autonomy. That is, a person is mentally disturbed to the degree that his internal mental states prevent him from making self-determined choices. We also see the attempt to embrace various psychological theories without being bound to one particular model.

Yet, in spite of the disavowal, the reader interested in these issues should go through the DSM-III to see whether it has really been successful in weeding out classical natural law and statistical deviancy criteria.

More directly, and related to psychiatric and other medical practice, the student (and any physician) should carefully examine the ways in which he himself categorizes, or tends to categorize, persons as "mentally disturbed," or their decisions as "irrational." Is he being biased by a unconscious view of how people ought to behave in society, or is there intellectual substantiation for his beliefs?

None of this is to say that the growing adoption of autonomy criteria for mental health has left us with no conceptual problems. It is easier to make general statements about autonomy than it is to provide criteria for determining when and if a person actually lacks autonomy and *is* being "irrational." And that conceptual problem presents itself constantly in the actual practice of psychiatry. It is the crucial issue in involuntary commitment, in the right to refuse psychiatric and medical treatment, and in criminal cases in which the defendant pleads "not guilty by reason of insanity." Consider some of the many ways in which the issue occurs: Is the Jehovah's Witness irrational when he refuses whole blood transfusions? Is the infertile couple justified in saying that having a child by artificial insemination is essential to their mental health? Is the pop singer irrational who takes female hormones in order to keep his voice high and his records selling? Is there a real difference between rational adherence to "a religion" and being "brainwashed" by "a cult"? Is the parent who believes that children should suffer severe physical punishment for any misbehavior irrational or guilty of child abuse?

INVOLUNTARY COMMITMENT

The fact that a person is mentally disturbed or psychotic no longer means that he is automatically committable to an institution. We may think of his situation as somewhat comparable to that of a person suffering from a somatic illness. We cannot forcibly bring persons with physical illnesses into treatment simply because they are ill. This was not true even in the recent past. Until recently, the natural law assumptions and precedents of law and psychiatry tended to make involuntary commitment rather simple. In the past two decades, however, the situation has been changing because of a number of factors: The development of the major psychotropic drugs gave numbers of nonfunctional patients the capability of functioning outside of institutions. Patients who would have spent years in institutions now could be medicated and released in a matter of weeks. In certain models of mental illness, they would still be considered psychotic, but compensated to the degree that the psychosis would not cause behavior that would necessitate institutionalization. Along with that, there was a reexamination of the practice of involuntary commitment in the light of the growing societal interest in the protection of civil liberties.

In the famous case of Donaldson v. O'Connor,[15]

strict criteria were set for involuntary commitment. Donaldson had been an inmate in a state institution for a number of years. After a number of attempts to obtain release, the courts finally ruled that he had to be released because he was neither a danger to himself nor a danger to others. State laws are still in the process of adjusting to Donaldson v. O'Connor and tend to allow involuntary commitment on only two grounds: (1) that the person is a danger to others because of his mental illness; and (2) that the person is a danger to himself because of his mental illness.

A number of ethical issues have arisen as a result of these changes.

The psychiatric profession welcomed the possibility of deinstitutionalization, assuming that the process would not be one of simply "dumping" patients out into society. The hope was that the combination of medication and a proliferation of outpatient care facilities would enable patients to do very well on the outside. However, the economic reality was that outpatient facilities were not set up and patients were, in fact, dumped. This has led to what has been called a "revolving-door" situation. Patients are involuntarily committed, and treated with the help of medication. No longer a danger to themselves or others, they are released. Once released, many face pressured economic situations. Many also stop taking medication, either because they believe they can do without it, or because they do not want to suffer the side effects. Some of these patients decompensate to the degree that they become dangerous to themselves or to others and are committed again, to restart the cycle. Still others decompensate, but not to a point at which it would be permissible to involuntarily commit them. The latter may live a miserable existence on the streets, or in borderline situations in society.

The ethical dilemma that has emerged is clear: Given that this group needs treatment. How do we provide them with treatment they do not want, without violating their civil liberties?

Danger to Self

There are also conceptual difficulties inherent in trying to determine what sort of behavior constitutes being "a danger to onself." While the determination may be easy to make if the person has made a direct and clear attempt at suicide, other cases are not that simple. In the case that opens this chapter, a claim was made that the woman's use of abusive language would possibly cause other persons to assault her and, hence, she was a danger to herself. Claims have been made that the homeless who choose to live in the street in winter rather than accept public shelters are suicidal and dangers to themselves. In the opening case, this argument was made even though the woman had already managed to survive at least two winters living in the streets. Counterclaims have also been made that a choice to live in the streets or in bus stations could be considered rational, considering the restrictive conditions and prevalent crime conditions that exist in some public shelters.

Could the danger-to-self criterion be used to commit cigarette smokers? Or individuals whose cholesterol intake is too high? It is also important to note that simply being a "danger to oneself" is no longer usually thought to be a sufficient reason for classifying a person as mentally disturbed. For example, our movement away from natural law views of illness has included a movement away from the belief that having a desire not to live is always sufficient evidence to warrant a diagnosis of mental disturbance.[16] This has been brought to the fore in cases of persons with terminal illnesses. We are seeing a growing acceptance of a belief that letting go of life may sometimes be a rational choice. The belief is still far from universal, and difficult cases relating to this issue and involving psychiatrists are common. We will speak about such cases when we deal with liaison psychiatry.

There are broader problems having to do with "danger to self": For example, there are persons with depressions that are presently untreatable. Some of these persons are now lobbying for the right to commit suicide. Analysis of such an issue is complex: Do we weigh the probability that treatments are forthcoming more heavily than we weigh a wish which may be rational within the context of present available treatments? Does society have the right to say "You must hold on"?

Danger to Others

In the first instance we should notice that committing someone because he is a danger to others has nothing directly to do with psychiatric treatment. Clearly this rationale for involuntary commitment is to use the powers of the state to promote a general societal interest, i.e., to protect society against the actions of an individual. Some psychiatrists will sometimes claim that to stop a person from harming others is really to the person's "best interests," for it will get him into treatment before he gets into trouble. But that claim is often nothing more than a rationalization for the psychiatrist's underlying utilitarian belief that societal interests weigh more heavily than individual rights. Sometimes too, the claim is a rationalization for a paternalistic belief that it is more important to treat the individual than it is to preserve his civil liberties.

Moreover, it has *not* been universally established that an involuntarily committed patient has a fundamental legal right to be treated, once he is in the institution. In fact, one of Donaldson's complaints to the court was that he had not received treatment while institutionalized. The court refused to rule on the justness of that complaint. States and municipalities may mandate that mental institutions in their jurisdictions provide treatment; but they are not yet required by federal law to provide treatment.[17]

Thus, some compare this kind of involuntary commitment to the preventive detention by the police of an individual who has not been found to be mentally disturbed, but is thought to be likely to commit a serious crime. Libertarians such as Szasz[18] believe that detention before a crime has been committed is a violation of the individual's liberty and his First and Fourth Amendment rights. In the libertarian view, no person ought to be detained before a crime has actually been committed or, at least, before the crime is in the process of happening. They believe the just policy would be to wait until a harm had actually been done and then to make an arrest. The question of the criminal's mental status could then become an issue in his trial or in his sentencing.

A utilitarian might support the libertarian policy on the grounds that a precedent of preventive detention could be used as a wedge to open up a general and classical totalitarian policy of preventive detention that would be used to imprison individuals at any possible threat of unorthodox behavior. That, in turn, will lead to general unhappiness in society.

While the libertarian point of view is somewhat consistent with autonomy theory, the autonomy theorist does differ in one important respect. He may say that if we can determine that a person is not autonomous, then his/her rights to liberty are at least temporarily abrogatable and he may, in fact, be institutionalized. That is, the right against preventive detention is a right held only by autonomous, competent moral agents—persons who have the ability to control their actions. A social contract theorist might offer the same argument. The implication of this approach is that any determination of a lack of autonomy and of endangerment would have to follow legal due process restrictions. That is, the individual would have to have the right to be represented by legal counsel, and the hearing would have to be judicial rather than simply a decision made by psychiatrists. For all that, another serious problem would remain—that of predicting that the person will, in fact, commit a violent act. Psychiatrists are in general agreement that, in most instances, they do not have the ability to predict violent acts with any reasonable probability. This problem is crucial to some of the issues relating to confidentiality, and is discussed more fully in Chapter 6 "Confidentiality."

There would still remain doubt about the status of the institution for involuntary commitment. As it stands now, if there is no guarantee of treatment for the person committed because he is a danger to others, then one might wonder if the place of commitment for those detained solely because of endangerment should be called a "hospital."

It is also important to note that just because someone has been determined to be a danger to others, it does not automatically follow that he must be mentally disturbed.

It was once a not unpopular point of view to hold that anyone who commits a crime must be at least somewhat disturbed.[19] Problems with that belief should be apparent. For example, someone may violate a law on grounds of principle, or someone may rationally gamble that the rewards of a crime outweigh the probability of being caught and convicted. There also are those who claim that their violation of a law was an intentional, rational, political act.

A less extreme position is the claim, already alluded to, that anyone who commits a *violent* crime must be severely disturbed. The underlying assumption here is that humans normally have an instinct against harming other members of their species. That assumption is dubious and, even if it were to be valid, the fact that we have a particular instinct doesn't itself entail that the instinct can't or shouldn't be voluntarily overcome by the individual.[20] Moreover, there may be times in which violence may be justifiable or, at least, times at which a reasonable case can be made for violence, such as in periods of war.

Other problems exist for those who believe that a case can be made for involuntary commitment on the grounds that the person is a danger to others. What precisely is meant by a "danger to others"? Can we commit the shopping-bag lady simply because she has lice and sits in crowded public buses? Does the infestation of lice constitute enough "danger to others" to warrant commitment? What about the addicted smoker who smokes in public places?

THE RIGHT TO REFUSE PSYCHIATRIC TREATMENT

It used to be true that an involuntarily committed mental patient retained very few rights. That is no longer the case, and among the now-recognized rights of these patients is the right to refuse treatment, including psychiatric treatment. We have already spoken about distinctions being made in regard to competency in Chapter 2, "Informed Consent and the Right to Refuse Treatment." Competency is now being seen as context related. That is, a finding that a person is incompetent in one area of functioning does not entail that he is incompetent in other areas. Thus, a person who is involuntarily committed is effectively being found "incompetent to go freely in society." However, that no longer means that he is incompetent to make decisions regarding his treatment. Psychotropic medications may have major side effects, and convulsive shock therapy can pose substantial risk to a patient. At least on grounds of those risks, such modalities require a separate informed consent. Because a patient is involuntarily committed does not mean that he hasn't the capability of making a competent decision regarding his treatment.

Other issues may arise when a patient requests a treatment modality that differs from the psychiatrist's concept of appropriate treatment. While this problem sometimes comes up as an issue in ordinary medical treatment, there are usually fairly tight boundaries that encompass appropriate alternative medical treatments. That is not as true in psychiatric care. The various psychological models may have different and conflicting prescriptions for treatment. A patient may feel that some other model of treatment is more appropriate for his own case, and a question may come up about the psychiatrist's obligations to stay with the patient in such cases. A psychiatrist dealing with an outpatient may be able to make a referral to another psychiatrist; he cannot solve the problem as easily with an inpatient.

Management and Treatment

The major tranquilizers have two effects. They are useful in treatment of psychoses. However, they also have the effect of calming persons down, whether or not the person is psychotic. That dual function leads to another kind of ethical issue that can show up in inpatient psychiatric care. There are times when patients "act up" in ways that are disruptive to the psychiatric unit and sometimes in ways that may endanger other patients or staff. These patient "management" problems may be dealt with in various ways: sometimes by secluding the patient, sometimes by using mechanical restraints, and sometimes by the administration of tranquilizers. There is a blurry line between *treating* a patient and *calming him down*. Nonetheless, there are times when an order is written for tranquilizers with the sole intent of *managing* a patient, but with the claimed purpose of *treating* him. Clearly, there may be times when management is necessary, and most hospitals have policies that require a rigorous procedure in order to guard against the misuse of "management" means for trivial purposes. There is the danger, however, that a "treatment" order will be written in order to bypass the stringent requirements for writing a "management" order.

LIAISON PSYCHIATRY

A psychiatrist who is called in for a consultation by a treating physician may face other sorts of ethical issues. Very often, the consult is called because the treating physician is faced with a patient who is refusing treatment. The attending physician who calls the consult may tacitly assume that any patient who refuses treatment must be irrational. Peer pressure is often brought to bear by the attending physicians in such cases. Their usual agenda is to call for a psychiatric consult for the express purpose of having the patient declared incompetent so that he may be treated without his consent. In fact and practice, a hasty psychiatric "diagnosis" of "incompetence" is often made. Very often the only criterion used to justify the judgment of incompetence is the patient's refusal of treatment. As we have said in Chapter 2, the fact that a patient refuses treatment that others think is appropriate is not alone sufficient to warrant a decision that the patient is incompetent.

FORENSIC PSYCHIATRIC ISSUES

Any physician who works in conjunction with the criminal justice system faces a number of controversial issues. That is particularly the case when the physician is a psychiatrist. Consider some of the things that forensic psychiatrists are asked to do. They may be asked to evaluate an arrestee to determine whether he is competent to stand trial. They may be asked to treat detainees to make them competent to stand trial. They may be asked to testify at trials where an insanity defense has been offered. They may be asked to treat prisoners and in some cases specifically those prisoners who have been found incompetent to be executed, in order to try to make them competent to be executed. In some states, they may be asked to help determine whether a prisoner is incapable of rehabilitation, and thus a candidate for execution.

Double Agency

Each of these tasks raises a number of ethical issues. For example, the forensic psychiatrist evaluating a prisoner for competency to stand trial may face a problem about confidentiality. Since he is evaluating the prisoner for the criminal justice system and not treating him, is he bound by the same duties of confidentiality as a treating psychiatrist? If not, does he have an obligation to warn the evaluee that he will not keep confidentiality? (cf. Chapter 6, "Confidentiality.") In treating detainees before they stand trial, the psychiatrist may face another sort of problem. If the prisoner intends to plead "not guilty by reason of insanity," his chances of winning his case are probably decreased if he appears in court after a period of successful treatment. That is, he is probably better off appearing "crazy" in court. Bound by his obligation to "do no harm," should the psychiatrist refrain from pretrial treatment so that his patient has a better chance in court?

The Insanity Defense

In testimony in cases where an insanity defense or related defenses of "diminished capacity" are offered, the forensic psychiatrist faces still another set of problems. First, there are conceptual problems tied to the insanity defense. Essentially, a plea of "not guilty by reason of insanity" is a claim that the accused was not capable of controlling his actions at the time of the crime.

That is a claim that is distinct from a claim that the accused had a mental disorder at the time of the crime. Proof that the defendant was psychotic at the time of the crime is not sufficient to show that he was unable to control his actions. The criteria used to make that judgment vary from jurisdiction to jurisdiction, and have also changed over time. While the verdict of guilt or innocence is decided by the jury or judge, the psychiatrist may have enormous influence on that decision. Yet the connections between mental disorders and loss of the ability to control one's actions are very much in dispute. This is an issue not only in the insanity defenses in lurid murder trials that become prominent in the newspapers but also in less lurid "diminished capacity defenses" based on diagnoses such as alcoholism, drug addiction, kleptomania, compulsive gambling, and premenstrual syndrome ("late luteal phase dysphoric disorder").

On a deeper conceptual level, the issue arises in conjunction with the implication of some psychological theories that no one every really has control over his actions, i.e., the belief that all actions, criminal or not, stem from causes that are out of the person's control, be they innate, conditioned, or psychodynamic. Indeed, it may sometimes seem that the criminal law, which assumes that persons are usually responsible and free to choose, is at odds with psychological theories which sometimes seem to imply that there are no freely chosen actions. As we have said, the acceptance of internal autonomy as at least a requisite for mental health clearly presumes that there can *be* free will. Nevertheless, there continues to be an ongoing conceptual tension between psychological theories tied to classical scientific deterministic underpinnings, on the one hand, and on the other, the law and those other conceptions of human essence, such as that of autonomy theory, that are tied to a belief in free will.

There are occasions in which psychiatrists are asked to take part in the punishment of prisoners. These occasions may range from treating convictees who have been given a choice of either going to prison or seeking psychiatric treatment (e.g., drivers convicted of drunken driving, and drug addicts) to treating prisoners on death row who have been determined to be incompetent to be executed. These again are instances in which there may appear to be a conflict between the physician's duties to his patient and his obligations towards society and his contractual obligations as an employee of an institution. Psychiatrists treating diverted offenders, such as in alcohol or drug treatment programs, often face patients who are not in treatment willingly and, in some instances, patients who are not compliant with treatment. To report such patients to a parole officer or to the courts may mean that the patient will be taken out of treatment and sent to prison.

A recent Supreme Court decision held that prisoners must be mentally competent to be executed. The rationale for the decision includes the belief that a convictee has a right to make last-effort appeals, which would require that he have the mental wherewithal to make appeals. In addition, there are the beliefs that a person must know the reasons for his punishment and also must understand and appreciate the fact that is he being punished. The decision has put some psychiatrists into a bizarre and, some think, morally untenable position. They are being asked to treat a person not with the intent of making him totally well, but instead with the intent of making him well enough to be killed. The code of ethics of the American Medical Association forbids physicians to take part in executions. The rule was likely aimed at requests that physicians administer lethal drugs to prisoners sentenced to death. In these cases, however, it is not clear that giving psychiatric treatment to these prisoners constitutes taking part in an execution. Some psychiatrists will refuse to treat such prisoners; some have called for the American Psychiatric Association to forbid any psychiatrist to treat such prisoners. Suffice it to say, the issue is a complex one. It takes in issues such as the meaning of "Do no harm," the obligations of a profession, the meaning and function of the practice of punishment, and the morality of capital punishment.

Child Custody Cases

Psychiatrists are often asked to testify in child custody cases, and sometimes asked to offer an opinion in adoption cases. They are often asked about the fitness of a parent to raise his child. Clearly, these are situations that are rife with ethical and conceptual issues. How does one determine what constitutes acceptable or desirable child raising "techniques," or determine who is psychologically fit or unfit to be a parent? These issues have come to the public attention recently in instances in which single persons wanted to adopt a child, and in which homosexual couples wanted to adopt a child or gain custody of a child after a divorce. It has also come up in at least one case in which the courts tried to take guardianship of children whose parents were members of a religious sect that believe that the Bible requires them to administer frequent corporal punishment to one's children; the court believed that the frequent punishment constituted child abuse.

Once again, we are confronted with ethical issues that are directly tied to the conceptual issue about defining "mental health" in an objective, value-free way.

Chapter 6

Confidentiality

- A patient enters the hospital complaining of a cough, and is diagnosed with *Pneumocystis carinii*. A diagnosis of AIDS is confirmed. He is married, and a well-known businessman in a small town. Upon learning of the diagnosis, he begs the physician not to tell his wife his diagnosis. His prognosis is poor, and he is not expected to survive his hospital stay.
- A patient tests positive for the presence of HIV antibody. She presents with no other significant medical problems. The patient is a female. She is a prostitute and she is an IV drug user. She lives with her boyfriend. Upon questioning, she claims that she is aware of the risks to her clients, but she says she needs the income, and she intends to continue working as a prostitute. In passing, she also hints that she would like to have a child

with her boyfriend. Questioned further about her intention to have a child, she admits uncertainty.

- A patient in an outpatient drug treatment facility has been tested positive for HIV antibodies. He has a history of antisocial behavior. His case record indicates that he has a steady girlfriend, and the case record contains her name and address. After counseling him about the ways of avoiding transmitting the virus, he is asked if he will tell his girlfriend about the positive test and if he will use condoms. He replies that he hasn't decided yet.
- During the course of a medical examination of an elderly patient on Medicaid, the patient reveals that he has a small amount of money in a bank account that would make him ineligible for Medicaid.
- A patient with a dire prognosis asks the physician not to tell his family. He has a wife and two children, one 17 years old and living at home, the other, 8 years old. They visit regularly and express great concern about the patient. They ask the physician about the condition of the patient.
- A patient presents with gonorrhea. He claims that he contracted it from a prostitute while on a business trip. He is married and has three minor children. He asks the physician not to report the case to the state or to his wife. He claims that he will refrain from intimate contact with his wife and anyone else until he is cured. He claims that his marriage is going through a difficult time and that if his wife were told it would mean the end of his marriage.
- A woman comes to the emergency room with numerous lacerations and a broken arm. She claims

that her husband beat her, and that this wasn't the first incident. She begs the physician not to tell anyone, saying she is afraid her husband will do worse.

- A woman has repeatedly shown up for treatment at a clinic during school hours with a 7-year-old child. In the course of conversation, she mentions that she keeps the child out of school a few days each week to help her around the house. When a suggestion is made that the child ought to be in school, she expresses anger and says "It's none of your business." The state has a law requiring the reporting of child abuse.
- A 14-year-old girl comes to see you with a request for birth-control pills. The state has a law requiring that minors bring proof that they have notified their parents before birth control devices can be prescribed. When asked, she says her parents would never agree.
- An executive of a large firm comes in for a yearly checkup which is required by her company. In taking a history, she reveals that she occasionally uses cocaine. Immediately afterward, she panics and asks you to remove the entry. She says if the firm knew that she uses drugs, she would lose her job.
- A patient undergoing outpatient treatment with a psychiatrist mentions that he had committed a homocide 8 years previously. He says that someone else was convicted of the crime and is now serving time.

The one horn of the ethical dilemma of confidentiality is the duty of the physician not to divulge information that his patient doesn't want divulged. The other horn has to be with the duty not to harm, and may point in various directions. Keeping confidentiality may result in harm to the patient himself, as in the case of the abused wife. Keeping confidentiality may result in future harm to some identifiable person other than the patient, as is the case with the prostitute and her boyfriend and the drug-treatment patient and his girlfriend. Keeping confidentiality may result in harm to some unidentifiable persons, as is the case with the HIV-positive prostitute. Keeping confidentiality may result in harm to a yet nonexisting person, as may be the case with the prostitute if she becomes pregnant. Keeping confidentiality may mean being complicit in ongoing harm to some person, as is the case with the patient who admits that someone else is serving time for a crime that he committed. Keeping confidentiality may result in some harm to society, as is the case with the patient on Medicaid.

The very question of whether keeping confidentiality will cause harm may be an issue, as in the case of the patient with AIDS who will die in the hospital, the case of the mother who keeps her child out of school, the case of the teenager asking for birth control pills, the case of the patient asking that his family members not be told his condition, or the case of the executive who uses cocaine.

The medical tradition of keeping patient confidences dates back at least to the school of Hippocrates. The Hippocratic Oath requires that the physician swear that "what I may see or hear in the course of treatment or even outside of the treatment in regard to the life of men, which on no account one must spread abroad, I will keep to myself holding such things shameful to be spoken about."

A cynical viewer of the Hippocratic injunction may

intepret the directive as self-serving rather than as a moral precept. He would hold that the Hippocratic physicians were interested in keeping powerful and rich patients, and wanted to encourage their trade by offering patients a guarantee that they would be immune from blackmail or embarrassment. We shall speak of less cynical justifications for the Hippocratic injunction below. However, whether or not the cynical view is correct, the Hippocratic injunction to keep confidences has come down to us as both a moral and legal obligation for physicians.

PHILOSOPHICAL JUSTIFICATIONS

Philosophical justifications for the obligation to keep confidences may be given on the grounds of various theories.

Utilitarian Theory

Society has an interest in encouraging people to seek health care. If persons know that what they say to their physicians will be held in confidence, they will be more likely to seek medical care. That is particularly true if the patients have disorders that might be thought to be embarrassing or stigmatizing, such as venereal diseases and psychological disturbances. Moreover, the physician on a house call becomes privy to all sorts of information about the patient and his household, much of which may be irrelevant to the direct care of the patient, and some of which might be embarrassing to the patient and his family if it were to be divulged. Patients would be more likely to seek care if they know that the physician will keep that sort of information private.

Furthermore, since only the physician has the expertise to decide what facts about the patient are relevant to treatment, it is important that the patient feel that he can be forthright about what he tells the physician. The more candid the patient is, the more information the physician will have, and the more information he has, the better he is able to treat.

Divulgence

Theoretically, utilitarian theory allows for the divulging of information if divulgence will produce a greater good or less harm for society than secrecy. However, as we have said before, it is often difficult to make accurate predictions about the utilitarian consequences of a policy or action. For example, as we will see, much of the present debate about confidentiality and HIV antibody testing stems from disagreements about the long-term consequences for the spread of AIDS of keeping or breaking confidentiality about test results.

Social Contract Theory

Sometimes a practice that has been traditionally and widely accepted can come to be formally protected by law. Because there has been a long-standing tradition of medical confidentiality, patients can be said to reasonably expect confidentiality when they go into treatment. Thus, it could be said that an obligation of confidentiality on the part of the physician is an implicit part of the special contract created when a physician takes on a person as his patient.

Autonomy Theory

Autonomy theory support for obligations to keep confidences could be given along the same lines as social contract theory. That is, because there is a tradition of confidentiality in the physician–patient relationship, the patient can reasonably expect that a physician will adhere to that tradition. For both theories, the obligation would be strengthened if the physician explicitly promised the patient that he would keep confidentiality. Moreover, those who believe that a recognition of persons' autonomy includes a recognition that persons have a right to privacy would also use that right as a basis for the right to confidentiality.

Divulgence

Kant himself seemed to believe than an explicit promise of confidentiality should never be broken, regardless of the consequences.[1] Modern versions of autonomy theory and contract theory tend to believe that breaking confidentiality is permissible, if not obligatory, in situations where keeping confidentiality would result in a substantial harm to others. The reasons that might be given to justify divulgence include: (1) No right is absolute; a right may be overridden if honoring that right will result in harm to others; and (2) Any person who enters a physician–patient relationship is expected to be aware of that limitation on his right to confidentiality. That is, just as no one can enter a theater expecting that he has the right to shout "Fire!" no one can go into treatment expecting that a physician will keep confidences when doing so will result in harm to the patient himself, or to other persons.

Some have argued that the duty to keep confidentiality is absolute, and that the physician–patient relationship should follow the model of the priest–penitent relationship in which the priest may reveal nothing that has been confided to him in the confession box. Arguments supporting this absolute view have been given on various grounds. There have been utilitarian arguments offered. These claim that the general benefits of absolute confidentiality are so enormous that any immediate benefit gained by allowing divulgence in individual cases would be outweighed by the long-term negative consequences of such a policy. Patients would not come into treatment or, if they did, they would not be forthright with their physicians if they knew that their physicians could decide to divulge information about them.

Classical Kantian arguments for an absolute duty of confidentiality have also been given. The claim here is that a person has an absolute right that any implicit or explicit promise made to him be kept.

It has also been suggested that requiring physicians to make decisions about divulging confidences would place upon them undue and unjustified burden. Physicians would be placed in the position of being policemen, having to make decisions that have nothing to do with the practice of medicine.

The more commonly accepted view is that the physician sometimes can and should break confidentiality without a patient's permission, but only under extraordinary circumstances. What those circumstances are is, of course, what the controversy is about.

It is important to note that neither social contract theory nor autonomy theory preclude the possibility of a physician and patient explicitly setting out limits to confidentiality, or even agreeing to a total waiver of confidentiality, at the onset of treatment. We will speak more about this later.

THE DEBATE ABOUT DISCLOSURE

There have always been claims that there are situations in which the physician not only *may* break confidentiality, but has a *duty* to break confidentiality.

For example, there may be a duty to disclose information about the medical condition of important public officeholders. Thus, the physician may have an obligation to divulge information about the health of the President even without the President's permission. That obligation might be explained in terms of social contract theory: In taking on the duties of public office, the President agrees to waive some of the rights that private citizens have. The practice of demanding such a waiver is justified by appealing to the public interest. It is important that the public have access to information about the President's health or, at least, access to information about medical conditions that might affect his ability to fulfill the responsibilities of his office.

The specific issue of confidentiality and public officials such as the President is unusual; however, there are analogous instances that are not so unusual and more controversial. For example, does a taxi driver who suffers from epilepsy have a right to confidentiality regarding his medical records? Does the medical student who was once hospitalized for depression have the right to confidentiality?

The more common claims to exceptions to the duty of confidentiality relate to situations in which a failure to divulge information will produce serious harm to other persons or to the patient himself. In these instances, the common view is that not only does the physician have the right to break confidentiality, but has a moral or legal obligation to do so.

The general belief is that the physician has a duty to disclose information if he can reasonably make the judgment that his patient poses a serious and imminent danger to himself or to others. The threat to others does not have to be intentional on the part of the patient. For example, the physician would have a duty to disclose information about infants who carry serious infectious diseases.

TARASOFF AND ITS IMPLICATIONS

Legal duties to divulge information may be discharged in various ways, depending on the context. In some instances, such as gunshot wounds, the physician may be required to inform the police that his patient has suffered a gunshot wound. In instances of child abuse, the physician may be required to inform the police or a child welfare agency. In other instances, such as infectious diseases, the physician may be required to inform a government agency. Also, in some instances, the physician may be required to inform the specific person who is in danger.

A legal case in California in the mid-1970s relating to confidentiality has had and continues to have enormous repercussions for the issue of medical confidentiality. The case is a good focal point for discussion of many of the issues involved in confidentiality.

The case was Tarasoff vs. Regents of University of California.[2] Prosenjit Poddar, a graduate student at the University of California at Berkeley developed an obsession for a woman, Tatania Tarasoff. Tarasoff barely knew him. Upon the suggestion of friends, he entered psychotherapy at a university clinic. During the course of therapy, he told his psychologist that he intended to kill Tarasoff. Tarasoff was on vacation in South America at that time. The psychologist consulted with his supervising psychiatrist, and a decision was made to notify the police. The police were notified, and they picked up and questioned Poddar. They decided that he was not dangerous and released him. Poddar did not return to treatment. When Tarasoff returned from South America, Poddar killed her. Tarasoff's parents brought suit against the therapist, claiming that he had a duty to directly warn Tarasoff that she was in danger. The parents won their case, and the Tarasoff decision has been used as the precedent for similar cases.

In the Tarasoff case, the court argued that while ordinary persons do not have a duty to directly warn the victims of intended violence, the "special relationship" that exists between the physician or therapist and his patient creates special duties to protect the patient or other persons from harmful actions of the patient. The special duty was first interpreted solely as a duty to warn identifiable probable victims. However, the courts later broadened the meaning of "protect." Under the later interpretation, a psychiatrist's obligation could be satisfied by any reasonable effort he made to protect probable victims whether or not he warned them. For example, his obligation to protect would probably be satisfied if he were able to arrange for the commitment of his patient.

Objections to Tarasoff

Tarasoff and similar decisions have caused great turmoil in the psychiatric community. Objections to the decision have been raised on a number of grounds.

First, psychiatrists have argued that they simply do not have the ability to predict the probability of violent behavior by their patients. Various empirical studies have been put forth to support that claim.[3]

Second, and connected to the first objection, psychiatrists claim that complete trust is crucial to the therapeutic relationship. Patients must be able to say anything, to give voice to any fantasy or feeling, without the fear that what they say will have consequences for them outside of the treatment room. Moreover, psychiatrists claim that patients often express their violent feelings and fantasies in therapy because they want to deal with those feelings as therapeutic issues. If confidentiality is broken, these patients will leave treatment. In some cases, violent feelings that could have been worked through in therapy will eventuate in action because the patient is no longer in treatment. Furthermore, psychiatrists, knowing they will be liable for a failure to warn, will tend to err on the side of overwarning. As a result, many patients will be lost from treatment.

Third, a false warning can have dire consequences for the patient. It may result in a temporary involuntary commitment, or the destruction of a relationship, or the loss of employment. A false warning may also have legal repercussions for the psychiatrist, as the patient may sue him for breaking confidentiality.

Fourth, there is too much ambiguity about the type of future behavior that warrants a duty to warn. How serious must the probable harm be in order to create a duty to warn? Should it include only possible murders? Property damage? Petty thefts? In a recent case in Vermont, the Tarasoff duty was extended to a case in which the patient's actions resulted in property damage without direct injury to any person. Does our case of the patient who admits that someone else is being unjustly punished for a crime he himself committed warrant a warning? Should a psychiatrist warn the transit company if a patient tells him that he intends to put graffiti on a bus?

Fifth, how far should the psychiatrist be expected to go in identifying and protecting probable victims? Suppose, for example, that Poddar's therapist knew only Tarasoff's first name, or only had a vague idea of her address? Should he be expected to make an effort to do "police work" in order to get more information? Suppose that Tarasoff had been told, but was unconvinced of the danger—should the psychiatrist have made an attempt to see her and convince her?

Sixth, psychiatrists claim that establishing a duty to warn will discourage persons from entering psychotherapy. The policy then will have the opposite effect of that intended; many violent persons who could have been dissuaded from violent behavior in the course of treatment will not enter treatment.

Replying to some of these objections, the court argued that the psychiatrist is not required to be totally accurate in his predictions, but is only required to exercise "that reasonable degree of skill, knowledge, and care ordinarily possessed and exercised by members of [that professional specialty] under similar circumstances."[4] The court also argued that the problems engendered by the tendency to overpredict are outweighed by the benefits of saving lives.

While the Tarasoff decision was limited to psychotherapeutic relationships, the court did base its opinion on that more general claim about the special relationship that exists between physicians and their patients. Thus, it is likely that Tarasoff will be used as a basis for extending to other medical situations the legal duty to warn when the patient poses a serious danger to other identifiable persons.

AIDS AND CONFIDENTIALITY

The appearance of the AIDS virus has raised a number of ethical issues relating to confidentiality.

Physicians have a legal obligation to report cases of certain serious infectious diseases to government agencies. However, in the case of reporting infectious diseases, some mechanism is put into place to preserve at least some of the patient's privacy. Information that would identify the patient cannot be made public. If there is a policy of "contact tracing" for the disease, workers from the public health services do not identify the patient to his possible contacts.[5]

The AIDS virus presents somewhat unique problems. First, much of the past legislation requiring mandatory testing and reporting of infectious diseases came at a time before there was a wide concern about protecting civil liberties and privacy. For many, a concern with protecting the public from AIDS has been tempered by a concern about the misuse of personal information by both the private sector and government agencies.

Second, because of public panic about AIDS, the disclosure that a person has AIDS or is HIV-positive can have serious consequences for that person. He may be ostracized, lose his employment or housing, be kept from public schools and, perhaps, be denied medical care.

Third, unlike most infectious diseases presently existing in our society, AIDS appears to be inevitably fatal and, as of now, is incurable. Thus, some may argue that while any duty to directly warn victims of diseases like gonorrhea is mitigated because the diseases are easily curable, the danger presented by AIDS strengthens the duty to warn. That is, under ordinary circumstances any harm that might follow from a failure to warn the spouse of a patient with gonorrhea is reparable. That is not so with AIDS. On the other hand, and contrary to that, unlike gonorrhea, AIDS is not easily transmittable. Thus, some argue that, in fact, difficulty of transmission of the AIDS virus mitigates the duty to warn.

Some have also offered a "let the buyer beware" argument against divulging information about a person's HIV status. They have argued that the main routes of transmission are known and avoidable. Persons should be expected to know that intravenous drug use and some forms of sexual behavior are risky, and therefore the onus of avoiding those dangers rests with them.

In turn, and counter to that, it has been argued that there are instances in which we wouldn't expect possible victims to assume that they are at risk. For example, we wouldn't expect a wife to be cognizant of her husband's secretive bisexual activity or secretive intravenous drug use. It has also been argued that we simply cannot depend on the supposition that probable victims are educated about risky behavior and, given the seriousness of AIDS, we should err on the side of assuming that they are not educated.

MANDATORY HIV-ANTIBODY TESTING AND DISCLOSURE

Ethical issues about mandatory testing for the presence of the HIV virus are closely tied to the issues about confidentiality. After all, a major purpose for mandatory testing would be to warn other persons about risks associated with those who test positive.[6]

Mandatory testing for the HIV virus has been recommended for many different contexts. Recommendations have ranged from proposals that all patients entering hospitals be tested to suggestions that we test all adults in the population. We shall speak here only about the issue of testing all patients entering hospitals.

The argument for testing all hospital patients for their HIV status is based mainly on a concern that HIV-positive patients pose a risk to health care providers. It has been argued that in certain areas, such as surgery and kidney dialysis, it is next to impossible for providers to take sufficient precautions against blood or fluid contact. Surgeons inevitably suffer scalpel cuts and needle punctures during surgery, and dialysis workers are constantly exposed to blood as the result of machine malfunctions. In other areas, such as in psychiatric wards, there may be a danger to providers and other patients because of violent behavior on the part of some patients, or there may be a danger to other patients who may have sexual contact with infected patients.

A counterargument to testing all patients is that testing will produce more harm than benefit. Patients will be reluctant to enter hospitals if they know they are to be tested, and the inevitable leak of information about HIV-positive patients will cause other patients to panic. The claim here is that it is better to take blood and fluid precautions with all patients. Counter to that, it is argued that while many hospitals do take some blood and fluid precautions with all patients, it is inconceivable that hospitals would have the wherewithal to take total blood and fluid precautions, such as isolation, with all patients. Another counterargument to mandatory and universal testing is based on the time presently needed to get a verified HIV blood test back from the laboratories. The final verification after an initial positive reading from a screening test may take weeks. In most cases, any risky procedure such as surgery would have to be done before the test results are back. Moreover, a negative test result would not ensure that the patient is HIV-negative, since the "gestation" period for the virus and antibody production may be months rather than hours.

In reply, those who favor mandatory testing say that testing will diminish the risk to providers by at least identifying some carriers. Those against mandatory testing reply in turn that, given that health care providers have an obligation to treat HIV-positive patients, and given that completely effective protective measures may be impossible in areas such as surgery, the negative consequences of a policy of mandatory testing would outweigh the positive consequences.

D.A. is a 32-year-old male with a clinical diagnosis of mixed substance dependence, including IV drug use and needle sharing.

During an admission workup, he was found to have a positive HIV. Mr. A. was counseled regarding the ramifications of the positive HIV finding, his responsibilities towards others. He spent 6 weeks in an inpatient drug rehabilitation program and was discharged after its completion.

When he returned for his outpatient weekly therapy program, he was accompanied by a girlfriend. The girlfriend waited outside during his therapy sessions. She was known by name and address to the treatment staff. During group therapy sessions, the patient freely admitted to an active sexual relationship with the woman. He said that he chose not to tell his girlfriend that he was HIV-positive because he feared that she would leave him. He said that he did not use condoms because they interfered with his pleasure.

If we take Tarasoff as a model, it is likely that the duty to warn upheld in that decision will be extended to establish a duty to warn identifiable probable victims of AIDS. The case above seems, at first glance, a clear instance in which there would be a duty to warn. Nevertheless, there are important questions at hand. One might ask, for example, whether an attempt was made to convince the patient to tell his girlfriend of his HIV status. It might also be asked whether the patient's avowed reluctance to tell her was due to passing and temporary psychological reactions to hearing his diagnosis. Was the patient experiencing denial and hostile reactions which are temporary and which can be worked out with therapy? If so, how much effort ought to be expended to help him work through those feelings before resorting to warning his girlfriend? We might also ask whether the patient knew the limits of confidentiality when he agreed to take the HIV test. Was he aware that confidentiality would be broken under certain circumstances?

GENERAL COMMENTS ON CONFIDENTIALITY

Given that there are instances in which there is a duty to warn a potential victim, some might argue that there should be a number of safeguards set up to avoid any misconception about and "misapplication" of that duty. For example, perhaps patients should be told at the onset of treatment that there are limits to their right to confidentiality. In the case of HIV testing, perhaps there should be a separate informed consent, as well as counseling about the risks to other persons of certain behaviors. In the case of those psychiatric patients who are able to control their behavior, they too should be warned about the limits of confidentiality when they enter treatment. Perhaps, too, a committee should review a case before a decision is made to divulge information.

There are, of course, difficulties with warning patients about the limits of confidentiality. In the case of psychiatric patients, the patient may develop a transference, and he may forget that his therapist is not to be totally trusted to keep silence. Such patients may require repeated warnings. However, repeated warnings, in turn, may interfere with treatment. Moreover, there is a problem about the specificity of any warning to patients about the limits of confidentiality. The details of the warning will have to depend on the type and seriousness of harmful behavior that warrants divulgence; and specifying that behavior, as we have said, presents a problem.

Warning patients who plan to take an HIV test about the limits of confidentiality may discourage some from taking the test. Many states do have independent test centers that preserve confidentiality. Offering that option to patients may alleviate a potential coercive situation in which a patient who wants to be tested for his own benefit has fears about the privacy of the results. Nevertheless, if issuing prior warnings about the limits of confidentiality in a hospital or therapy setting does have the consequence of discouraging patients from being tested, that will likely have negative effects on our ability to gather important statistics about AIDS. It may also discourage patients for whom the test would be important for their own future treatment in the hospital or therapy setting.

Yet if we do believe that there are instances in which

there is a duty to break confidentiality, it would seem wrong to allow patients to accept the HIV test or go into therapy with an assumption of complete confidentiality, only later to surprise them with a break of that confidentiality.

Chapter 7
Afterword

We have only touched on the range of medical ethical issues. The time allocated for formal exposure to medical ethics in medical school is usually quite limited. Given that limitation, we have tried to cover the "nuts and bolts" issues that are of most immediate concern in clinical work. That is not to denigrate the importance of the other issues, nor to say that they will not come up as immediate concerns in clinical work. We can only say a few words about some of those issues here.

ALLOCATION OF SCARCE RESOURCES

While issues relating to the scarcity of available organs for transplant may not come up in the early clinical years, other allocation issues may present themselves. For example, there may be a problem of too many patients, too few beds, and too few treatment staff in an intensive care unit. The allocation issues have two aspects. They present themselves as "micro" issues which the clinician has to deal with immediately in his work. They also present themselves as "macro" social issues. How should society go about procuring more organs for transplant? Should public monies be spent in finding ways to procure more organs, in research in areas like growing organs, or in preventative health measures to alleviate the initial need for organs?

ACCESS TO HEALTH CARE

Related to allocation issues are questions pertaining to access to health care. Does the government have a responsibility to provide more hospital beds and more staff, or should the health care system work as a free market system with the number of beds and staff to be determined by profit considerations? Does society have an obligation to attempt to provide all needy persons with organs (or, in fact, any health care) if needed? How has the development of reimbursement according to diagnostic related groups (DRGs) affected health care in general and the obligations of physicians in particular?

The application of analytical philosophical thinking and an examination of the implications of the various philosophical theories can be very useful in thinking through solutions to the larger issues. Each of the theories has implications for issues such as the allocation of scarce resources and the distribution and access to health care. Some relevant reading sources are presented in the Bibliography.

RESEARCH AND EXPERIMENTATION

We have not focused on research issues, although we have mentioned some of them in passing. Much of what has been presented in Chapter 2 can be applied to some of the research issues regarding consent. There are other issues which have been only touched on or not mentioned at all. These include problems about the use of experimental procedures on patients, the experimental use of fetuses, children, and incompetent adults, and the use of animals in research and education. Here too, the reader is referred to the Bibliography.

GENETIC ENGINEERING, SCREENING, AND REPRODUCTIVE TECHNOLOGY

Our increasing ability to diagnose and predict the occurrence of genetically linked disorders and our growing ability to examine and alter genetic structures have led to concern about ethical implications. Some of the ethical issues are not new. For example, there has long been a concern about involuntary sterilization and mandatory screening. Other questions, such as those related to our abilities to fertilize ova in vitro and transplant fertilized ova, are new. What has been written here in regard to informed consent, personhood, and mandatory testing for the presence of HIV antibodies can serve as a basis of discussion. There has been increased attention to these challenges, and a large body of literature has developed.

The large social issues are somewhat different in nature from the clinical ethical issues. Many of the ethical clinical issues arise because the larger social issue has not been resolved. Thus, if there were sufficient available transplant organs, or sufficient hospital beds, there would be fewer clinical ethical issues about triage. If DRG reimbursement limits were a real indication of the hospital time needed for total patient care, there would be no problems of conflict between a physician's desire to treat fully and a hospital's desire to be cost-effective and profitable. If there were more outpatient psychiatric units, there would be fewer of the "revolving door" problems discussed in Chapter 5, "Psychiatric Ethics." This double aspect to the issues makes resolution of particular case questions especially difficult, for resolution requires activity on two levels. The physician qua physician has to work hard trying to find equitable resolutions on the micro level. That is, for example, given that there are too few beds in the intensive care unit, he has to try to allocate those beds in a fair way. The physician qua citizen has to work on a civic level, using the evidence he has gleaned from his clinical experience and thinking hard, carefully, and analytically about the micro problems. As a citizen who confronts these issues every day, he must work to put just and fair policies into effect in society at large. The latter task may appear to be both hopeless and thankless. Nevertheless, in spite of the changing nature of physicians' status in society, physicians still have a powerful voice. A physician's letter to a medical journal, to a congressman, or to a newspaper can have an influence on the framing of policy. Changes in codes of ethics, in laws, and in hospital policies often have their start from the discussion initiated by such letters. A case brought to a hospital ethics committee (and a demand by physicians for an active ethics committee in their institution) can initiate change within a hospital.

It is hoped that the material presented in this book will be of some use for those who do wish to approach the issues in a rational and coherent manner.

Notes

CHAPTER 1

1. More strictly, we should say that any interaction involving sentient beings brings in questions about our obligations to those beings. The issue about our obligations to nonhuman animals is a large and important one, but cannot be included in the limited space in this book. See bibliography for books that deal with this issue.
2. This will be discussed in more detail below in the section on "Rights and Obligations."
3. Sometimes the grumbling is really an expression of a sort of legal positivism ("We will do what the law requires us to do"). C.f. the section on rights in this chapter.
4. The claim was popularized by certain anthropologists, in particular, Margaret Mead. A part of a recent controversy about her work was about the validity of her claim that moral beliefs are culture-relative.
5. The terms "moral" and "ethical" will be used interchangeably in this book.
6. See following section, "Utilitarian Theory."
7. Sometimes, justification is given based on evolutionary theory.

That is, the belief that "survival of the fittest" implies a competition among individuals in a species. But it should be noted that even classical evolutionary theory makes claims not only about survival of individuals, but also about survival of the species. Sometimes individual survival seems to be sacrificed for the survival of the species or other individuals. (The author begs the indulgence of those sophisticated in both ethics and evolutionary theory. The objection here is admittedly oversimplified.)

8. Immanuel Kant, *Foundations of the Metaphysics of Morals.* Trans. by Lewis White Beck. (Indianapolis: Bobbs-Merrill, 1959).
9. The ascription of a "will," not the sort of thing that can be observed, is the sort of thing that empiricists would avoid.
10. Kant, op. cit.
11. Bruno Bettelheim, *The Informed Heart* (London, 1961), Chapter 6. As quoted in Jonathan Glover, *Causing Death and Saving Lives* (New York: Penguin, 1977) p. 58.
12. Thus, a dog can be taught to walk on two legs, but because the dog is "designed" to walk on fours it will be neither healthy nor happy if it continually walks on two legs.
13. This is reportage, not recommendation. We note that the Pope recently (January 1984) reiterated the classical Church natural law position on sexual matters—sex outside of marriage is a "moral disorder," masturbation and homosexuality are also "grave moral disorders." Notice the interesting concept of a "moral disorder"—is it a disorder in the medical sense—an illness?
14. Thus, the Catholic Church has been quite "permissive" on withdrawing certain life-sustaining treatments on the grounds that the use of extraordinary means is not obligatory since extraordinary means are "above and beyond" the natural order.
15. This may not hold true for the physician who may have special duties arising from his role as a professional and arising from his contract with the patient.
16. Thomas Hobbes, *Leviathan* (Parts I and II) (Indianapolis: Bobbs-Merrill, 1958).
17. Hobbes, p. 108.
18. Ibid., p. 107.
19. On the one hand, this was a very radical notion (and Hobbes got in a great deal of trouble for it)—for it claims that government is put into place by the governed and has obligations towards

them. This was written during the time of absolute monarchies. On the other hand, it is conservative in the sense that it justifies a minimalist government—one that has an obligation to protect the liberties of the governed, but no obligation to do anything more for them.

20. The reader might find it interesting to reexamine the section on "self-interest" above.
21. John Rawls, *A Theory of Justice* (Cambridge, Mass: Belknap Press, 1971). Rawls presents a very full discussion of this view.
22. In regard to medical ethics, the last becomes important when we deal with the issue of drug-testing and drug sales of untested or dangerous drugs to other countries.
23. In theory. In conservative practice there may be inconsistencies, for example, in the demand that aggressive treatment be given to defective neonates. (See Chapter 4, "Euthanasia," and see section on abortion in Chapter 3, "Personhood and the Right to Life.")
24. Though there have been attempts to claim that these rights are derivative negative rights because good health, nutrition, etc., are necessary in order for any person to have a real opportunity to exercise his primary rights to pursue liberty, happiness, etc.
25. Cf. Rawls, p. 25.

CHAPTER 2

1. Though these medical instances are usually treated as "civil" batteries rather than as "criminal" batteries. In some jurisdictions the legal concept of "battery" has been subsumed under the legal concept of "assault."
2. Part of the intent in classifying a failure to get consent as a battery is to establish a basis for the awarding of damages. That is, there is a long precedent for making awards in battery cases. But that doesn't capture the whole story either.
3. Schloendorff v. New York Hospital. 211 N.Y. 127,129: 105 N.E. 92,93 (1914).
4. *NY Times* 6/25/87 pB12.
5. At this point it is unclear what, if any, legal sanctions there would be against the physician who refused to offer such an explanation after his patient asked him for an explanation.

6. Yet there are difficulties implied by that exemption. For example, the physician should be aware that what he thinks is a commonly known procedure may not be commonly known and understood by the layperson. We find the same difficulty in explaining procedures to a patient—every trade and profession has its jargon, and medicine is no exception. The use of some terminology can become so second-nature to the physician that he may easily forget that the terminology is peculiar to his profession rather than commonly used in society.
7. Probably, a further implication is made that there would be little risk of harm to the patient from the procedure if he "happened" to consent without having known what he had consented to.
8. Although what follows from that refusal in regard to the physician's or hospital's obligation to continue treatment of the patient is a problem in and of itself. See below, "The Right to Refuse Treatment."
9. American Hospital Association, *A Patient's Bill of Rights*, (American Hospital Association, 1972).
10. It appears that some physicians can now refer their patients to a telephone service "Med-line" which will give the patients a tape recording of information about particular diseases or procedures.
11. See Chapter 5, "Psychiatric Ethics."
12. A fuller discussion appears in Chapter 5, "Psychiatric Ethics."
13. We are only examining surface claims here. The claim has also been made that the underlying intent of such legislation is really to construct a legal basis towards an eventual overturning of Roe vs. Wade.
14. Superintendent of Belchertown State School v. Saikewicz, 373 Mass. 728, 370 N.E.2d 417, 424 (1977).
15. Two criteria to which, unfortunately, many do not adhere. Sometimes a resident will ask a student to do something beyond his capacity or beyond his right at that stage of his education, be it a procedure, or a direction to obtain an informed consent.
16. It is hoped and assumed that the preceptor will not insist that his students lie about their status to patients the next time.
17. American Medical Association, *Current Opinions of the Judicial Council of the American Medical Association* (Chicago: American Medical Association, 1982), pp. 28–29.
18. The issue of choices available to poor and well-off patients will be examined in Chapter 7.

CHAPTER 3

1. From this point, we will use the term "fetus" to refer to unborn humans at all stages of development. We generally will not distinguish the term fetus from terms such as "conceptus," "zygote," and "embryo."
2. There are other factors, too, that include issues of inheritance, as well as issues of criminal charges of homocide. For example, a married couple with siblings but no children are shot while being robbed. They have no written wills. Both sustain severe injuries. After 2 days the wife is brain dead but kept somatically alive. The husband suffers irreversible cardiac arrest on the third day. Given that inheritances go to the blood relatives of the last surviving spouse, which siblings should get the inheritance? Moreover, is the assailant to be charged with two homicides or one?
3. A Definition of Irreversible Coma, Report of the Ad Hoc Committee of the Harvard Medical School. *Journal of the American Medical Association*, Vol 205, No. 6 (August 1968), pp. 337–340.
4. President's Commission for the Study of Ethical Problems in Medicine and Biomedical and Behavioral Research. *Defining Death*, (Washington, D.C.: U.S. Government Printing Office, 1971).
5. Ibid., p. 2.
6. Thus, for example, the parents of a child who had suffered head trauma resulting in irreversible coma in a state in which the brain-death criteria had not been accepted requested that the child's life support be removed. When the courts refused, they attempted to have the child moved to a state in which he could be declared dead.
7. In Europe, and in some states, dye injections may be used to determine whether or not there is brain-cell activity. Sometimes too, dye tests may be used to confirm an electroencephalograph reading.
8. There are circumstances in which life-support interventions will be continued. These include cases in which permission has been given to harvest organs, unusual cases in which the "neomort" is pregnant and there has been a request or legal order to try to bring the fetus to viability, and, sometimes, instances in which the patient or his family had religious objections to brain-death criteria for death.

9. Anne Rot and H.A.H. van Till, Neocortical Death after Cardiac Arrest. *Lancet*, (November 1971), p. 1142.
10. Remembering that under most whole-brain-death criteria, including the Harvard criteria, the presence of any of these activities does not allow a declaration of death.
11. Ronald Melzack, *The Puzzle of Pain* (New York: Basic Books, 1973), pp. 15–17.
12. Not entirely, of course.
13. Though some would argue about that "certainty."
14. Of course, that may be irrelevant to the formation of a just policy. Likely, most laypersons would have a hard time seeing the brain-dead person as a dead person.
15. J. David Bleich, Minority Opinion, in The New York State Task Force on Life and the Law, *Do Not Resuscitate Orders* (New York: New York State Task Force on Life and the Law, 1986).
16. This term is borrowed from Michael Tooley. Abortion and Infanticide, *Philosophy and Public Affairs*, 2 (Fall 1972), pp. 37–65.
17. Plato, *Theaitetos*. Trans. John Warrington (London: J. M. Dent and Sons, 1961), pp. 67–157.
18. I have depended heavily on *The Morality of Abortion* by John Noonan (Cambridge, Mass: Harvard University Press, 1970) for my discussion of the history of abortion in Catholic thought.
19. Roe v. Wade. 410 U.S. 113, 93 S.CT. 705 (1973).
20. Ibid.
21. The reader should also refer to Chapter 1.
22. Although there might be a question about the possibility of making a Kantian universal law permitting abortion.
23. See Chapter 1 for an explanation of "perfect" and "imperfect" duties.
24. An argument might be given to claim that they have a positive right to life also.
25. For philosophers, we are simplifying matters by treating theories such as that by John Locke as mixtures of natural law and contract theories.
26. Cf. Jay E. Kantor. The Interests of Natural Objects. *Environmental Ethics*, 2 (Summer 1980), p. 168., Michael Tooley, op cit.
27. More will be said about the positive right to life later.
28. Cf. Kantor, Tooley, op cit.
29. What is sometimes called "reflective" consciousness.
30. This is a confused claim made by St. Thomas and Hobbes, among others, who spoke of having innate desires to live. Or-

ganisms, including humans, may have innate responses to stimuli, and those responses may tend to keep the organism alive, but that is different from having a concept and desire to stay alive per se. Thus, for example, the kitten or newborn infant that avoids falling off high places almost certainly is not thinking to itself, "If I crawl any further, I will fall off and die." Clearly, any claims about the degrees or lack of degrees of nonhuman animal consciousness are controversial.

31. This raises an important issue of animal rights. This argument may entail that earthworms have a right not to be put into needless pain even though they may not have a right to life. Some books on animal rights are referred to in the Bibliography.
32. And we really face similar difficulties when we try to determine the degree of consciousness of nonhuman primates which seem to have complex brains and complex behaviors.
33. Remembering that a right to life entails that violation of the right will mean inquest, trial, and possibly, severe punishment.
34. We will discuss this issue in Chapter 5, "Psychiatric Ethics."
35. See Chapter 1.
36. At this writing, there are court cases considering this very issue.
37. There are a number of cases in which women have been forced to undergo procedures such as cesarean sections for the sakes of their fetuses. These cases are in appeal at the time that this is written.
38. Though we should reiterate that classical theories of rights also seem to imply that infants do not have rights.

CHAPTER 4

1. Cf. James Rachels, Active and Passive Euthanasia, *New England Journal of Medicine,* 292 (January 1975), pp. 78–80.
2. More correctly, we should distinguish the legal concept of "competence" from the psychiatric concept of "capacity." "Competency" is a legal rather than a medical or philosophical term, and is really decided by the courts. We will use the term "competent" here for the sake of simplicity. *See* Chapters 2 and 5 for a fuller discussion.
3. e.g., a terminally ill patient is in unmanageable pain. He requires constant attention by staff, and an enormous amount of ex-

penditure of money and hospital resources. Other patients with better prognoses are suffering as a result of the time, attention, and resources spent on this patient. He prefers to hang on. However, the staff decides that everyone (including the patient) would be better off if he were dead.

4. See section on "Voluntary Passive Euthanasia" in this chapter.
5. e.g., *Current Opinions of the Judicial Council of the American Medical Association*, pp. 9–10.
6. Discussion of Roe v. Wade in Chapter 3.
7. Kant argued strongly against suicide, partly on the grounds that it was an abandonment of one's duties towards one's own person. However, Kant wrote before our modern medical technology, and it would seem compatible with Kant to say that a person who was falling into a state in which he could no longer perform his duties to himself or others could not be said to be "abandoning" his duties.
8. The reader is referred to Chapters 1 and 4 for a fuller discussion of "extraordinary" and "ordinary" means.
9. In certain cases, it may not even make sense to speak about either "harm" or "benefit" to the patient. It is certainly doubtful that a "brain-dead" patient could be either harmed or benefited by anything that is done to him. It is probably doubtful that patients who are in persistent vegetative states could be either harmed or benefitted. c.f. Chapter 3.
10. In re Conroy, No. A-108 (N.J. Sup. Ct. Jan. 17, 1985).
11. e.g., President's Commission for the Study of Ethical Problems in Medicine and Biomedical and Behavioral Research, *Deciding to Forego Life-Sustaining Treatment* (Washington, D.C: U.S. Government Printing Office, 1983), esp. pp. 87–89.
12. The assumption here is that there both are and ought to be regular discussions with the patient about his treatment.
13. The first sense is also beginning to be used by persons in regard to their treatment wishes if and when they were to become incompetent as a result of psychosis. This is the so-called "Ulysses Contract," after the episode in the *Odyssey* in which Odysseus (Ulysses), knowing he would be enthralled by the song of the Sirens, but wanting to hear them, ordered his men to tie him to the mast and ignore any command that he might give while under the Sirens' spell. The concept was first proposed by Thomas Szasz. Thomas S. Szasz. The Psychiatric Will—A New Mechanism for Protecting Persons against 'Psychosis' and Psychiatry. *American Psychologist*, 37 (July 1982), 767–770.

14. Cf. Chapter 2.
15. Society for the Right to Die. *The Physician and the Hopelessly Ill Patient*, (New York: Society for the Right to Die, 1985), p. 86.
16. This section should be read in conjunction with Chapter 2, "Informed Consent and the Right to Refuse Treatment."
17. Cf. Chapter 3. "Personhood and the Right to Life."
18. e.g., *Deciding to Forego Life-Sustaining Treatment*, op. cit. pp. 275–299.
19. Referral to an ethics committee may be mandatory in some hospitals.
20. In re Quinlan, 70 N.J. 10, 355 A. 2d 647 (1976).
21. case reference.
22. See Chapter 5, "Psychiatric Ethics."
23. The reader should also refer to Chapter 2 and Chapter 5.
24. "Quality of life" may be more applicable to those cases in which "endurance" is no longer an issue. That is, the brain-dead patient and the patient in deep irreversible coma may be said to have *no* quality of life.
25. It is important for the reader to read this section in conjunction with Chapter 3, "Personhood and the Right to Life."
26. "Defective neonate" is taken here to include very premature infants and infants with severe congenital problems as well as those with severe genetic defects.
27. For reasons that were mainly procedural, having to do with the lack of hearings prior to the directive, as well as a ruling that the directive was "arbitrary."
28. These considerations are, in fact, primary in a utilitarian analysis. However, it would seem that, at the least, the contractual relationship of patient–physician weighs heavily in favor of placing the obligation towards the patient as first.
29. Cf. Chapter 3, "Personhood and the Right to Life."
30. Cf. Chapter 2, "Informed Consent and the Right to Refuse Treatment."
31. Cf. Chapter 3, "Personhood and the Right to Life."
32. *Current Opinions of the Judicial Council of the American Medical Association*, op. cit., p. 9.
33. e.g., The Hemlock Society.
34. An elderly retired nurse suffering from a terminal illness requested her physician to give her a lethal dose of drugs. After discussion with the patient and her brother, the physician gave her the drugs and turned himself in to the police, presumably

to begin a test case. The physician was eventually found innocent.

35. Although courts are sometimes lenient in their treatment of family members convicted of active euthanasia.

CHAPTER 5

1. That is, to psychiatry, clinical psychology, and to some aspects of social work. Since this book is geared to physicians, we will use the term "psychiatry" throughout.
2. We will generally not distinguish among the terms "mental disorder," "mental disease," and "mental illness" in our discussions.
3. Although Aristotle, the "founder" of natural law theory, claimed to have derived his ideas about the essence of normal humanhood from his empirical observations of statistically normal behavior.
4. cf. Chapter 1.
5. The foremost proponent of this point of view is the psychiatrist, Dr. Thomas Szasz. Thomas S. Szasz, *The Myth of Mental Illness* (New York: Hoeber-Harper, 1961).
6. Consider, is infertility a disease? If so, is the physician who does a tubal ligation or vasectomy making his patient ill?
7. Plato, *The Republic*, e.g., 61.
8. For example, Darwinian theory discards the natural law notion that living species were designed with constant and unchangeable purposes, and replaces it with the idea that species can change as a result of environmental changes.
9. Jean Jacques Rousseau, *The Social Contract*, revised and edited Charles Frankle. (New York: Hafner, 1947).
10. For example, Skinnerian behaviorism and some versions of Freudian theory.
11. Immanuel Kant, *Anthropology from a Pragmatic Point of View*, Trans. V. L. Dowdell (Carbondale: Southern Illinois University Press, 1978).
12. We are not claiming that there is a direct causal connection between the acceptance of autonomy ethics and the changes in psychiatric theory. More likely, it is a zeitgeist phenomenon.
13. Perhaps the most dramatic recent example of this change was

the decision of the American Psychiatric Association to remove the diagnosis of homosexuality from its nosology of mental disorders.

14. *Diagnostic and Statistical Manual of Mental Disorders* (3rd edition) (Washington, D.C.: American Psychiatric Association, 1987), p. xxii.
15. O'Connor v. Donaldson, 422 U.S. 563 (1975). In fact, as we shall see, a general right to treatment for the involuntarily committed has never been established as a legal right in this country. We shall have more to say about the case when we speak of the right to refuse treatment and the right to treatment.
16. Natural law theorists usually held to the belief that we have an innate desire to stay alive.
17. Wyatt v. Stickney, 325F. Supp. 781 (M.D. Ala. 1971).
18. Szasz, op. cit.
19. We shall speak about this more when we deal with the insanity defense and the problem of free will.
20. Kantian autonomy theory is based on the belief that instincts can be willfully overcome.

CHAPTER 6

1. Immanuel Kant, On a Supposed Right to Lie from Altruistic Motives, in *Critique of Practical Reason and Other Writings in Moral Philosophy*. Trans. and Ed., Lewis White Beck (Chicago: University of Chicago Press, 1949), pp. 346–350.
2. *Tarasoff v. Regents of University of California*, 13 Cal 3d 177, 118 Cal Rptr 129,529 P2d (1974). *Tarasoff v. Regents of University of California*, 17 Cal 3d 425,131 Cal Rptr 14,551 P2d 334 (1976).
3. Cf. Tarasoff v. Board of Regents, op. cit., also Livermore, Malmquist, and Miehl. On the Justifications for Civil Commitment. *University of Pennsylvania Law Review*, Vol 19, p. 84 (1968).
4. Tarasoff v. Board of Regents, op. cit.
5. Of course, this may be more a theoretical protection of confidentiality than a practical protection. No doubt the majority of persons contacted can identify the person who is the source of risk.
6. There are other rationales too, including the need to gather statistical data about the prevalence of the virus, the need to protect the blood and other organ supply, and the need to notify persons who are tested that they are at risk.

Bibliography

CHAPTER 1: INTRODUCTION AND PHILOSOPHICAL THEORIES

Dworkin, Ronald. *Taking Rights Seriously*. Cambridge, Mass.: Harvard University Press, 1977.

Haworth, Lawrence. *Autonomy*. New Haven: Yale University Press, 1986.

Hobbes, Thomas. *Leviathan* (Parts I and II). Indianapolis: Bobbs-Merrill, 1958.

Kant, Immanuel. *Foundations of the Metaphysics of Morals*. Trans. Lewis White Beck. Indianapolis: Bobbs-Merrill, 1959.

McCloskey, H.J. Rights. *The Philosophical Quarterly*, 15, 1965: p. 123.

Melden, A. I. (Ed.). *Human Rights*. Belmont: Wadsworth, 1970.

Mill, John Stuart. *Essential Works of John Stuart Mill*. Ed. Max Lerner. New York: Bantam, 1961.

Nozick, Robert. *Anarchy, State, and Utopia*. New York: Basic Books, 1974.

Rawls, John. *A Theory of Justice*. Cambridge, Mass.: Belknap Press, 1971.

CHAPTER 2: INFORMED CONSENT AND THE RIGHT TO REFUSE TREATMENT

American Medical Association. *Current Opinions of the Judicial Council of the American Medical Association*. Chicago: American Medical Association, 1982.

Culver, Charles M. and Bernard Gert. Basic Ethical Concepts in Neurologic Practice. *Seminars in Neurology*, 4 Mar. 1984: 1–8.

Gibbs, Richard F. Informed Consent—What It Is and How to Obtain It. *Legal Aspects of Medical Practice*, 15 Aug, 1987: 1–6.

Superintendent of Belchertown State School v. Saikewicz. 373 Mass. 728, 370 N.E.2d 417, 424 (1977).

CHAPTER 3: PERSONHOOD AND THE RIGHT TO LIFE

Ad Hoc Committee of the Harvard Medical School to Examine the Definition of Brain Death. A Definition of Irreversible Coma. *Journal of the American Medical Association*, Vol 205, No. 6, 1968: 337.

Bleich, J. David. Minority Opinion. In The New York State Task Force on Life and the Law, *Do Not Resuscitate Orders*. New York: New York State Task Force on Life and the Law, 1986.

Feinberg, Joel (Ed.). *The Problem of Abortion*. (2nd ed.). Belmont: Wadsworth, 1984.

Kantor, Jay E. Some Rights of Some Non-Moral Agents—Necessary and Sufficient Conditions for Personhood. Diss. CUNY, 1978.

Kantor, Jay E. The Interests of Natural Objects. *Environmental Ethics*, 2 Summer, 1980.

Kindregan, Charles P. *Abortion, The Law, and Defective Children*. Washington, D.C.: Corpus, 1969.

Lamb, David. *Death, Brain Death, and Ethics*. New York: SUNY, 1985.

Melzack, Ronald. *The Puzzle of Pain*. New York: Basic Books, 1973.

Noonan, John. *The Morality of Abortion*. Cambridge, Mass: Harvard University Press, 1970.

Plato, *Theaitetos*. Trans. John Warrington. London: J.M. Dent and Sons, 1961.

President's Commission for the Study of Ethical Problems in Medicine and Biomedical and Behavioral Research. *Defining Death*. Washington, D.C.: U.S. Government Printing Office, 1971.

Roe v. Wade. 410 U.S. 113, 93 S.CT. 705 (1973).
Rott, Anne and H.A.H. van Till, Neocortical Death after Cardiac Arrest. *Lancet*, November, 1971.
Tooley, Michael. Abortion and Infanticide. *Philosophy and Public Affairs*, 2 Fall, 1972: 37–65.

CHAPTER 4: EUTHANASIA

Beauchamp, Tom and Seymour Perlin. *Ethical Issues in Death and Dying*. Englewood Cliffs: Prentice-Hall, 1978.
In re Quinlan, 70 N.J. 10, 355 A. 2d 647 (1976).
Ladd, John (Ed.). *Ethical Issues Relating to Life and Death*. New York: Oxford University Press, 1979.
President's Commission for the Study of Ethical Problems in Medicine and Biomedical and Behavioral Research. *Deciding to Forego Life-Sustaining Treatment*. Washington, D.C.: U.S. Government Printing Office, 1983.
Society for the Right to Die. *The Physician and the Hopelessly Ill Patient*. New York: Society for the Right to Die, 1985.
Steinbock, Bonnie. *Killing and Letting Die*. Englewood Cliffs: Prentice-Hall, 1980.
Szasz, Thomas A. The Psychiatric Will—A New Mechanism for Protecting Persons against "Psychosis" and Psychiatry. *American Psychologist*, 37 July, 1982: 767–770.
The Hastings Center. *Guidelines on the Termination of Life-Sustaining Treatment and the Care of the Dying*. Briarcliff Manor, N.Y.: The Hastings Center, 1987.

CHAPTER 5: PSYCHIATRIC ETHICS

American Psychiatric Association. *Diagnostic and Statistical Manual of Mental Disorders*. (3rd ed.). Washington, D.C.: American Psychiatric Association, 1987.
Culver, Charles M. and Bernard Gert. *Philosophy in Medicine*. New York: Oxford University Press, 1982.
Ennis, Bruce and Loren Siegel. *The Rights of Mental Patients*. New York: Avon, 1973.
Kant, Immanuel. *Anthropology from a Pragmatic Point of View*. Trans.

V.L. Dowdell. Carbondale: Southern Illinois University Press, 1978.
Moore, Michael S. *Law and Psychiatry*. New York: Cambridge University Press, 1984.
O'Connor v. Donaldson, 422 U.S. 563 1975.
Rosner, Richard and Harold J. Schwartz (Eds.). *Geriatric Psychiatry and the Law*. New York: Plenum, 1987.
Rousseau, Jean-Jacques. *The Social Contract*. Revised, trans. and ed. Charles Frankle. New York: Hafner, 1947.
Standing Committee on Association Standards for Criminal Justice of the American Bar Association. *Proposed Criminal Justice Mental Health Standards*. (Annual Meeting, 1984) American Bar Association.
Stone, Alan. *Law, Psychiatry and Morality*. Washington, D.C.: American Psychiatric Press, 1984.
Szasz, Thomas S. *Law, Liberty, and Psychiatry*. New York: Collier, 1963.
Szasz, Thomas S. *The Myth of Mental Illness*. New York: Hoeber-Harper, 1961.
Wyatt v. Stickney, 325F. Supp. 781 M.D. Ala. 1971.

CHAPTER 6: CONFIDENTIALITY

Kant, Immanuel. On a Supposed Right to Lie From Altruistic Motives. In *Critique of Practical Reason and Other Writings in Moral Philosophy*. Trans. and Ed. Lewis White Beck. Chicago: University of Chicago Press, 1949.
Livermore, Malmquist, and Miehl, On the Justifications for Civil Commitment. *University of Pennsylvania Law Review*, 19, 1968: 84.
Tarasoff v. Regents of University of California, 13 Cal 3d 177, 118 Cal Rptr 129,529 P2d 1974. Tarasoff v. Regents of University of California, 17 Cal 3d 425,131 Cal Rptr 14,551 P2d 334 1976.

CHAPTER 7: AFTERWORD

Bayles, Michael D. *Reproductive Ethics*. Englewood Cliffs: Prentice-Hall, 1984.

President's Commission for the Study of Ethical Problems and Biomedical and Behavioral Research. *Implementing Human Research Regulations*. Washington, D.C.: U.S. Government Printing Office, 1983.

President's Commission for the Study of Ethical Problems and Biomedical and Behavioral Research. *Splicing Life*. Washington, D.C.: U.S. Government Printing Office, 1982.

Regan, Tom and Peter Singer (Eds.). *Animal Rights and Human Obligations*. Englewood Cliffs: Prentice-Hall, 1976.

Singer, Peter. *Animal Liberation*. New York: New York Review of Books, 1975.

GENERAL SUGGESTED READINGS

Abrams, Natalie and Michael D. Buckner (Eds.). *Medical Ethics*. Cambridge, Mass.: MIT Press, 1983.

Beauchamp, Tom L. and LeRoy Walters (Eds.). *Contemporary Issues in Bioethics*. (2nd ed.). Belmont: Wadsworth, 1982.

Gorovitz, Samuel, et al. (Eds.). *Moral Problems in Medicine*. Englewood Cliffs: Prentice-Hall, 1976.

Harris, John. *The Value of Life*. London: Routledge & Kegan Paul, 1985.

Humber, James M., and Robert F. Almeder (Eds.). *Biomedical Ethics and the Law*. (2nd ed.). New York: Plenum, 1979.

Jonsen, Albert R., et al. *Clinical Ethics*. New York: Macmillan, 1982.

Mappes, Thomas A. and Jane S. Zembaty (Eds.). *Biomedical Ethics*. (2nd ed.). New York: McGraw-Hill, 1986.

Munson, Ronald (Ed.). *Intervention and Reflection: Basic Issues in Medical Ethics*. Belmont: Wadsworth, 1979.

O'Neill, Onora and William Ruddick (Eds.). *Having Children*. New York: Oxford University Press, 1979.

The Hastings Center. *The Hastings Center Report*. (bimonthly).

VanDeVeer, Donald, and Tom Regan. *Health Care Ethics*. Philadelphia: Temple University Press, 1987.

Wertz, Richard W. *Readings on Ethical and Social Issues in Biomedicine*. Englewood Cliffs: Prentice-Hall, 1973.

Index